Table of Contents

The National Juvenile Firesetter/Arson Control and Prevention Program

Fire Service Guide to a Juvenile Firesetter
Early Intervention Program

November, 1993

United States Fire Administration /Federal Emergency Management Agency

This project was supported by grant number 87-JS-CX-K104 awarded to the Institute for Social Analysis by the Office of Juvenile Justice and Delinquency Prevention and the U.S. Fire Administration. Points of view or opinions stated in this document are those of the authors and do not necessarily represent the official positions or policies of the Office of Juvenile Justice and Delinquency Prevention or the U.S. Fire Administration.

CHAPTER 1: INTRODUCTION

In Passaic, New Jersey, a firefighter was killed and hundreds of people lost their homes in a fire started by a group of teenage boys. In Roanoke, Virginia, a seven year old boy set fire to a chair in an abandoned building. The fire spread to an adjacent house and trapped an elderly woman. In Rochester, New York, a two year old, playing with matches, started a fire that took his life and the lives of five family members. Unfortunately, these tragic events are not isolated incidents, but are repeated virtually every day in cities and towns across the United States. Arson fires kill hundreds of people every year and cause over one *billion* dollars worth of damage annually.

It is estimated that approximately 40% of all arsons are set by juveniles. These fires cause hundreds of millions of dollars in damages annually and thousands of needless injuries and deaths. The rate of juvenile fireplay and firesetting -- short of arson as determined by fire investigators -- may also be quite high. Studies have shown that the majority of normal children possess an interest in tire and nearly half have engaged in fireplay. While the majority of the child set fires are set out of curiosity not malice, the damage they cause, both in economic and human costs, are real and devastating.

Clearly the problem of juvenile firesetting and arson is a costly, often deadly, problem. Whether the result of a curious child playing with matches or the malicious act of a troubled delinquent, juvenile firesetting is a serious and vexing problem that requires a special response from the community.

In recognition of the seriousness of the juvenile firesetter problem nationally, the Institute for Social Analysis was contracted by the Office of Juvenile Justice and Delinquency Prevention and the U.S. Fire Administration to conduct the National Juvenile Firesetter/Arson Control and Prevention Program (NJP/ACPP). This developmental initiative is designed to assess, develop, test, and disseminate information about promising approaches for the control and prevention of juvenile firesetting and arson.

ISA initially planned to develop one "model" program. However, we quickly realized that a program suited to a small, volunteer fire department in rural Minnesota would not be appropriate for a large, paid department in the Bronx and vice versa. Instead, ISA developed a modular or components model. The components describe how to develop, implement, and operate a juvenile firesetter program. These components highlight seven different aspects of a program, including program management, screening and evaluation, intervention services, referral mechanisms, publicity and outreach, monitoring systems, and developing relationships with the justice system. ISA, with assistance from its subcontractors and consultants developed three program manuals. The *Guidelines for Implementation* contains a broad range of information about each of the components. The *User 's Guide* provides step-by-step guidelines on how to use the information in the *Guidelines* to implement a juvenile firesetter program. The *Trainer 's Guide* describes how to use the information in the *Guidelines* as well as other resources to develop a comprehensive juvenile firesetter prevention training workshop. The guide provides a detailed curriculum for the training program and provides strategies for modifying the curriculum to meet specific needs

As noted above, the *Guidelines for Implementation* contains a vast range of information. It includes information not only for the fire service, but also for mental health, juvenile justice and child protective services agencies. In writing that manual, ISA attempted to provide the reader with the widest variety of program information and purposely did not specify the type of program a jurisdiction should implement. However, we have found that an "evaluation, education, and referral program" model is the one most often implemented by juvenile firesetter programs across the country. After the National Juvenile Firesetter/Arson Control and Prevention Program was evaluated, it became clear that a shorter manual specifically devoted to the development of an early intervention program was needed.

In writing this manual we have culled the information from the *Guidelines for Implementation* and presented what we believe to be the most effective program for the fire service to implement. Not every recommendation will pertain to all jurisdictions, and program planners should refer back to the *Guidelines* for most complete information on alternative approaches. Like the other program materials, this manual is written primarily for the fire service -- the agency most likely to house a firesetter program.

ISA advocates an "evaluation, education, and referral program" -- where the fire service provides initial screening and evaluation of firesetters who have been identified, early intervention/fire safety education for curiosity firesetters or other firesetters who may benefit from such education, and referral to mental health or other appropriate agencies for more troubled firesetters. Although some programs include a "counseling" component within the fire service, ISA has found that the fire service is best equipped to screen juvenile firesetters, generally using the screening tools developed by the U.S. Fire Administration, and provide fire safety education when appropriate. Any additional counseling is usually best handled by mental health or child protective services agencies.

ISA believes that the fire service should not have to tackle the problem of juvenile firesetting alone. Other agencies including schools, mental health agencies, juvenile justice agencies, child protective services, and other agencies that work with youth need to work with the fire service to help identify juvenile firesetters and provide the appropriate services so that these juveniles do not continue to set fires. In most communities, the initial impetus to develop a juvenile firesetter program comes from the fire service, but as this volume indicates, the fire service should solicit assistance from these other agencies -- often in the context of developing a coordinating council -- to provide all the necessary services.

This manual follows the format of the other materials developed under this program. It will discuss an early intervention and referral program in the context of the seven components ISA feels are critical to the success of any juvenile firesetter program. These components include 1) Program Management, 2) Screening and Evaluation, 3) Intervention Services, 4) Publicity and Outreach, 5) Referral Mechanisms, 6) Monitoring Systems and 7) Relationships with the Juvenile Justice System.

CHAPTER 2: PROGRAM MANAGEMENT

Program Planning

Juvenile firesetter programs often originate out of the concern of one inspired individual. These individuals may be members of the fire service who have a genuine interest in children or have seen, first hand, the damage and pain caused by juvenile firesetting. (If a program is established, they often serve as program coordinators). Once the interest is generated, the next step is to acquire a detailed understanding of the juvenile firesetter problem in the particular jurisdiction. Local fire departments are the first place to go to obtain the necessary information. Fire incident reports, arson investigation reports, and other records should provide data on the extent and nature of the juvenile firesetter problem. Information on property loss, injuries and deaths, if available, give added meaning to the numbers. Additional information and arrest data is often available from hospital burn units and police departments.

Once data on the extent of the local juvenile firesetter problem is collected, the person interested in the problem will need to meet with fire chiefs, community representatives, local city councils, and others to determine if the magnitude of the juvenile firesetting problem constitute a serious enough issue to warrant community action. These meetings often center around: (1) the cost of the problem versus the cost of the solution, (2) whether fires set by juveniles are a significant proportion of the fires set in the community and (3) whether the numbers of fires set by juveniles is disproportionally high given the numbers of juveniles in the community. Deciding whether the issue of juvenile firesetting is a significant problem will probably take many meetings and discussions. Do not be discourage if the process takes longer than expected. Many juvenile firesetter programs have been created because one individual had the tenacity to keep going back and presenting his or her case.

If juvenile firesetting is not considered a problem, the path is easy. If, however, the community decides that juvenile firesetting is a threat to the community, the next issue is deciding how to address the problem. The community may decide that the problem is severe enough to establish a complete juvenile firesetter program. Communities wishing to implement such a program can use the guidelines described in these components and can seek assistance from other juvenile firesetter programs. If a separate juvenile firesetter program is beyond the resources of a particular community, the tire service may choose to bolster existing programs by implementing one or more of the components. Each jurisdiction is unique and has their own unique problems and resources. Only the members of that community can decide what constitutes a serious problem and which strategies will be most effective to address the problem.

Program Structure

If the members of the community feel that the juvenile firesetter program is serious enough to warrant a juvenile firesetter program, they must decide what kind of program would best serve their jurisdictional needs and then develop such a program. As indicated earlier, ISA advocates an early intervention program directed by the fire service for single jurisdictions (we will briefly discuss the special case of multiple jurisdiction programs later). The following sections are designed to provide information on how to establish a juvenile firesetter program. The remainder of this component provides information on such issues as location, staffing, training, funding, and establishing interagency links.

Location

Fire Service. The primary site for a juvenile firesetter program should be within the fire service. The results of the Institute for Social Analysis' (ISA) survey of juvenile firesetter programs throughout the country reveal that 87% of the programs are administered by the fire service. These programs are located in different branches of the fire service, including the Office of the Fire Chief, Fire Investigation, and the Fire Marshal. The primary reason why juvenile firesetter programs should be established within the fire service is the fire service's capacity to identify large numbers of firesetters. The fire service is usually the first agency to respond to a fire and many of the firesetters are identified at the scene. Indeed, the majority of juvenile firesetter referrals to existing programs are from within the fire service, usually followed by parents and then schools and mental health organizations.

The fire service's knowledge of fire cause and origin facilitates their ability to identify youthful firesetters. The fire service can track a case from the identification of the firesetter through the fire investigation, assessment, and intervention (education, counseling, prosecution, etc.). In addition, many fire departments have established links with some of the crucial referral agencies. In the course of their investigations, fire investigators often communicate with police, probation, social service, and justice personnel. These links are vitally important to the success of a juvenile firesetter program (see the Referral Mechanisms Component).

Although the overwhelming majority of juvenile firesetter programs are housed within the fire service, there is still some concern that such a program will fall prey to departmental politics. The greatest concern is that the program would be terminated when the Fire Chief who instituted it leaves. Most departments feel that a juvenile firesetter program is too valuable to have such a tenuous existence.

Another concern is that firefighters and investigators may be hesitant to use the juvenile firesetter program. To overcome these potential problems, juvenile firesetter programs must be institutionalized within the fire service. Their existence can not rely solely on the motivation and drive of one individual. To survive, the juvenile firesetter program must receive support from all levels within the fire service and community.

To gain the support of all the fire service personnel, the program director should brief the chief and all the division heads about the juvenile firesetter program. Brief memos can be circulated to each fire service division describing the juvenile firesetter program services. Each firefighter, fire investigator, fire educator, etc. should know about the program and understand how it works. The Publicity and Outreach component will describe how to inform the general public about the juvenile firesetter program.

Multi-jurisdictional Approach

The majority of the juvenile firesetter programs surveyed by ISA functioned at the local level. More recently, however, a number of programs are considering a county or larger, multi-jurisdictional approach. Indeed, during the implementation and evaluation phase of the National Juvenile Firesetter/Arson Control and Prevention Program, the three sites selected for implementation all chose to implement a multi-jurisdictional approach. One of the greatest advantages of such an approach is that many of the referral agencies that work with the juvenile firesetter program (e.g., mental health, probation, juvenile court, etc.) are county, not local, agencies. A multi-jurisdictional program may span many towns and allow these communities to combine their resources instead of competing for the limited resources of county agencies with whom they work.

The multi-jurisdictional approach, however, requires additional planning and coordinating. The national evaluation of the NJF/ACPP, conducted by the American Institutes for Research (AIR), concluded that while "the fire service agency may still be the proper location for screening, education, and referral services for individual juveniles, ... it may not be the best choice for achieving regional coordination." Using the experiences of the three implementation sites, AIR suggests "An agency that already spans the boundaries of the region and that already has experience building and maintaining networks that may be a more effective program vehicle than an individual fire department" for managing a regional approach

In one of the implementation sites, the program management was operated out of the Council of Government's office. The Council had exactly the type of networking experiences necessary to implement a regional juvenile firesetter program. The Council had extensive contacts over the four county area and was able to bring over twenty fire departments together. As AIR points out, the Council also had a proven track record with the area's public officials. In another site, a non-profit organization worked with the District Attorney's Office to develop a firesetter program that encompassed numerous fire districts within one judicial district. The non-profit agency was able to take advantage of the District Attorney's district-wide network in bringing together the appropriate agencies. One advantage of using a non-fire service agency to manage a multi-jurisdictional juvenile firesetter program is that you can avoid inter-departmental rivalries that can sometimes impede the progress of a program.

If a site chooses to develop a multi-jurisdictional program, it will need to establish detailed guidelines for referral and feedback to each fire department/station within the juvenile firesetter program jurisdiction. Each fire department/station within the jurisdiction served by the program should be briefed about the program's purpose and services. As is true for local juvenile firesetter programs, the multi-jurisdictional program will be responsible for assessment, intervention, referral, case tracking and follow-up.

The feasibility of a multi-jurisdictional program will be based on the extent of the juvenile firesetter problem in a given jurisdiction and the resources available to ameliorate the problem. In some cases, the structure of the fire service or referral agencies may make such an approach untenable. Each jurisdiction will need to assess which approach is most appropriate to meet the needs of the community.

Staffing and Responsibilities

Fire Service Personnel. Programs located within the fire service should be coordinated by an individual with a genuine interest in the juvenile firesetter issue. Ideally that individual should be a senior ranking fire official. As noted earlier, programs need support from the highest level in the fire service. Many programs are administered by the Office of the Fire Chief or Fire Marshal with the coordinators answering directly to the Fire Chief or Fire Marshal. The coordinators would be responsible for the day-to-day activities of the juvenile firesetter program. They would be in charge of assessment and intervention, either directly or by supervising others who are assigned to provide the assessment and intervention services. The coordinators would also be primarily responsible for facilitating communication between the juvenile firesetter program and other agencies. The coordinators should be viewed as managers who are responsible for not only the mechanics of running the juvenile firesetter program, but also for the leadership and direction of the program.

In larger departments, additional fire service staff should also be assigned to the program. Firefighters, fire investigators, and fire educators can provide screening and evaluation services, fire education, and referral services. Fire service personnel can be trained to screen juvenile firesetters using standard screening instruments. In addition, fire service personnel can provide fire safety education to juvenile firesetters. Many fire departments employ fire educators who have the responsibility to teach fire safety in the schools. Providing fire safety education to the juvenile firesetters is often seen as an extension of that responsibility.

Staffing Issues and Concerns

The staffing structure presented above can be implemented by jurisdictions that have the manpower resources available to staff a juvenile firesetter program. Smaller jurisdictions, however, may not have the resources or the need to establish an actual juvenile firesetter program. These jurisdictions may simply want to incorporate some of the services provided by a juvenile firesetter program into existing agencies. Although it is recommended that jurisdictions establish a "program" within the fire department, an alternative possibility for some jurisdictions is to have interested people in various agencies take on the responsibilities of the juvenile firesetter program staff. These individuals would be responsible for assessment, education, referral, and tracking.

Jurisdictions must also decide whether the juvenile firesetter program will be operated by full-time or part-time staff and whether the staff will be paid or volunteer. Ideally, a juvenile firesetter program should be staffed by at least one full-time, paid fire service employee. Many fire departments have separate budgets for the juvenile firesetter program, which includes full-time personnel. In other jurisdictions, fire service personnel provide juvenile firesetter assessment, education, and referral in addition to other responsibilities, such as, fire investigations, fire inspections, or school fire safety education. Still other jurisdictions rely entirely on fire service personnel who volunteer their off-duty time to help provide assessment and education to juvenile firesetters. (In later sections, this component will discuss liability concerns surrounding the practice of having paid fire service staff volunteer their time to the juvenile firesetter program).

Training

Regardless of the staff background, all program staff should receive training in juvenile firesetting and child related issues. At a minimum, the training should include the following topics:

> Characteristics of juvenile firesetters
> How to identify juvenile firesetters
> Developing and managing a juvenile firesetter program
> Screening/assessment techniques
> Interviewing and educating the juvenile firesetter
> Referral and follow-up
> Normal child development
> Juvenile delinquency
> Child Abuse/Neglect
> Legal Issues

National experts in the field of juvenile firesetting can provide training in the characteristics of juvenile firesetters and information on how to identify, assess, interview, educate, and refer firesetters. Local fire service personnel can provide specific information about the jurisdiction's juvenile firesetter problem.

Personnel from local social services agencies and mental health facilities can provide training in child related issues, such as child development, delinquency, abuse, and neglect. Training in these child related issues is important to understanding juvenile firesetters. For example, in some cases, more seriously disturbed firesetters engage in other acts of juvenile delinquency. In other cases, youth will set a fire to draw attention to parental abuse or neglect. Juvenile firesetters represent a broad spectrum of youth, from developmentally normal children who are simply curious about fire to very seriously disturbed youth who require specialized treatment. The juvenile firesetter staff need this diverse training because they will come in contact with a wide range of juveniles in the course of their work.

Information about the legal issues surrounding juvenile firesetting can be obtained through the local prosecutor's office. Program staff need to be aware of the arson laws, including the age of accountability. The staff should also know how juvenile firesetters are handled by the justice system

Once the juvenile firesetter program has been established and the program staff have received training, the program coordinator or other staff should provide an orientation to all fire service personnel, especially arson investigators and upper level command staff. This may be done in the from of an in-service meeting or one day seminar. (If the resources are available, these personnel may also be included in the training seminar). All fire service personnel should be aware of the program and the services it provides. In addition, all fire service personnel should understand the procedures used to refer a firesetter to the program. Questions and concerns of the fire service personnel about the program should be addressed at this time. An example of a Juvenile Firesetter Training Workshop can be found in the *Trainer's Guide.*

In addition to the fire service orientation, the program coordinator should prepare briefings for the Chief and Deputy Chief to enhance their understanding of the problem and gain their support. As the program continues, the coordinator should provide the Chief and Deputy Chief with brief updates on the progress of the juvenile firesetter program.

Funding

As noted earlier, the nature and extent of any juvenile firesetter program will depend, to a large extent on the resources available to the program. Programs with limited money and manpower have gone to the community to acquire the necessary services, materials, and funds. The community can offer an unlimited wealth of resources. Corporations may contribute money or sponsor specific activities or products. When looking for corporate donations, juvenile firesetter program staff should appeal to the corporation's sense of civic mindedness and self-interest. Contributing money to better the community and help eliminate a costly and deadly problem is basically good business. The juvenile firesetter program should consider establishing local public/private partnerships. These partnerships, which include representatives from local businesses and public agencies (such as the fire service), have been useful in other government programs. Local businesses can donate more than money or equipment -- they can contribute their management, fund-raising expertise, and other in-kind contributions.

One potential resource for the juvenile firesetter program is local insurance companies. Where they may not always be able to offer monetary contributions, they may be able to provide in-kind assistance. Several juvenile firesetter programs have received generous help from the insurance industry regarding public relations activities, such as, printing brochures or publishing an article on the problem of juvenile firesetting and the promising program solutions. Insurance agencies can be a valuable resource because they have a vested interest in facilitating the reduction of juvenile firesetting.

Prior to approaching insurance agencies or other local businesses, the juvenile firesetter program should gather as much statistical information as possible about the juvenile firesetter problem in their community. Information about the cost of juvenile firesetting in economic and human terms will help support funding efforts. The most important element of juvenile firesetting that needs to be stressed is that the problem is a *community* problem that cannot be alleviated without the assistance of the community.

One additional source of funding can be investigated by the juvenile firesetter program. In some jurisdictions the restitution paid by a juvenile firesetter as part of a court sentence is not claimed by the insurance industry. Often this is because the cost of claiming that money is more than the actual amount of the restitution. The program coordinator should talk to the court and the insurance agencies about the possibility of earmarking those funds for the juvenile firesetter program.

Liability

Another financial (and legal) concern is the issue of liability. Liability refers to the potential for programs or referral agencies to be "at risk" for legal action because of the actions of a juvenile firesetter. Program staff need to take steps to insure that referral agencies will not be held liable for the actions of the juveniles referred to them. Liability waiver forms are often used to counter these concerns. The liability waivers should be reviewed by attorneys to make sure they address all the concerns of the juvenile firesetter program and the referral agencies. Parents will need to read and sign these waivers (which usually release the program or the referral agencies from responsibility for the action of their children). The juvenile firesetter program and the program's referral agencies need to be able to address the needs of the firesetters but will be limited if they are going to be held accountable for the actions of the firesetter.

Another liability issue arises when fire service personnel volunteer their time to work with juvenile firesetters. As mentioned earlier, paid fire service personnel sometimes volunteer their time to assist juvenile firesetters. The new Fair Labor Standards Act (FLSA) may limit this practice. Many states are interpreting the FLSA to mean that the fire service is liable for fire service personnel when they are conducting fire service related activities regardless of whether they are on- or off-duty. Juvenile firesetter programs that use volunteers from the fire service as part of their program need to carefully review their state's interpretation of the FLSA and how it affects the program.

How to Establish Interagency Links

Regardless of how the fire service chooses to tackle the juvenile firesetter problem, they will need the assistance of the key community agencies which work with juveniles (e.g., police, probation, justice, schools, mental health, and social services). The following section will describe techniques for gaining support from these community agencies and establishing interagency relationships

Coordinating Council. Strong interagency relationships and referral networks are vital when establishing a juvenile firesetter program. Because these relationships are so critical to the success of a program, the creation of a coordinating council is essential. Such coordinating councils or task forces have been established in Portland, Oregon; Upper Arlington, Ohio; Central Oklahoma, and other programs around the country. The juvenile firesetter program coordinating council should be composed of representatives from all agencies in the jurisdiction whose responsibilities relate to juvenile firesetters. At a minimum, the council should include representatives from the fire service, police, probation, juvenile court, children's protective services, district attorney's office, schools, and mental health agencies. These agencies represent the avenues through which juveniles are referred to the juvenile firesetter program, as well as, resources for the program. Including all of the key agencies on the coordinating council will ensure that no juvenile falls through the cracks and that all firesetters are identified, evaluated, and receive appropriate interventions. Ideally, the council should meet once a month to discuss the problems or concerns and develop future plans for the program.

The coordinator of the juvenile firesetter program will have the primary responsibility for recruiting the council representatives. If the coordinator is a member of the fire service, s/he may also represent the fire service on the council. The coordinator should contact the administrator of each agency to explain the juvenile firesetter program and the role of the council. The coordinator should then set a time to meet with the potential council members. The coordinator may want to have background materials, such as, statistics on the local juvenile firesetter problem and examples of how each agency is affected. It is important to stress that the problem of juvenile firesetting is a community problem that touches every agency mandated to provide services for juveniles. One program coordinator caught the attention of other agencies -- and ultimately won their support -- by telling them that the next child to die in a fire was their responsibility, not his. Descriptions of how other programs work may also help convince agencies that juvenile firesetter programs work, if they receive the support of other agencies. If the head of the agency is unable to participate on the council, s/he should suggest a representative, preferably the person most likely to have contact with juvenile firesetters.

Role of the Council. The primary role of the juvenile firesetter program coordinating council is to facilitate multi-agency cooperation to plan, implement, and maintain the juvenile firesetter program. The coordinating council should institute procedures for referrals to and from the juvenile firesetter program and should define the roles of each agency. For example, the juvenile firesetter program would be chiefly responsible for providing assessment and education, while child protective services and mental health agencies would provide counseling services for more troubled firesetters. Each agency representative could work toward providing the necessary procedures for acquiring services for juveniles referred by the juvenile firesetter program, such as, sliding fees, if necessary.

The council will be responsible for developing specific referral agreements between the juvenile firesetter program and different agencies. These referral mechanisms, which will be describe in detail later in the Referral Mechanisms Component, should include procedures for information exchange between the program and the referral agencies. Dual waivers and contracts enable the program staff to learn the status of the juveniles they refer for additional services. Referral agencies should also be able to learn the status of juveniles referred to the program.

One of the most important functions of the council representatives is to educate the other council members about their agency's strengths and limitations. Misunderstandings and problems between agencies often develop because one agency is not familiar with how the other agency operates. The tire service, for example, is designed for immediate response, but social service organizations are often not able to respond with the same speed. The workshop described below is designed to help personnel in different agencies understand how their counterparts work. The council will maintain communication between agencies and troubleshoot when necessary. If a firesetter is not receiving the services recommended by the juvenile firesetter program, the council or appropriate representatives can intervene to find out why.

Finally, council representatives will also be called upon to help identify other agencies or individuals who work with juvenile firesetters. Council representatives should disseminate information about the juvenile firesetter program to their agencies and the community and promote the program. The goal of the council is to gain support for program from all agencies that work with juveniles and to ensure that all those who work with juvenile firesetters understand the function of the juvenile firesetter program.

Juvenile Firesetter Program Workshop

As noted earlier, different agencies have different working cultures. To help agencies learn about each other, the juvenile firesetter program should sponsor a one-day workshop for employees of each of the agencies represented on the council. During the workshop, which is based on a seminar sponsored by Rochester, New York's FRY program, the members of the coordinating council will serve as a panel to moderate the workshop. Members of the key agencies who work with juvenile firesetters should be invited to attend. The workshop will give participants the opportunity to meet their counterparts in different agencies and learn how different agencies operate.

The juvenile firesetter program may want to have the State Fire Marshal or other representative give opening remarks or a keynote address. After the welcoming remarks, the program coordinator can begin the workshop by describing the characteristics of juvenile firesetters, the nature and extent of the problem in that jurisdiction, and the role of the juvenile firesetter program. The panel representatives can then be asked to describe how their agency works with juvenile firesetters. Attendees can be asked to share their experiences with juvenile firesetters. Participants should be encouraged to ask the panel members how they might handle a particular case.

After a break for lunch, attendees should be assigned to different groups. Each group should include at least one member from each of the different agencies attending the workshop. Each group should be given a description of a juvenile firesetting case and instructed to discuss how they would handle the case. Each group member would then be responsible for describing how his/her agency handles such cases. Participating in this activity gives each attendee the opportunity to see how different agencies handle the same case. Understanding different work styles and philosophies is essential if agencies are going to be asked to work together to solve the juvenile firesetter problem. The workshop can close with a discussion of the role of the juvenile firesetter program coordinating council and what the program needs from each agency.

The workshop should be conducted after the juvenile firesetter program has been established and the staff has been trained. It is designed to give the referral and resource agency staff a formal opportunity to learn about the juvenile firesetter program and meet their counterparts in other agencies.

Jurisdictional Characteristics

The structure of any jurisdiction's juvenile firesetter program will be affected by: 1) the size and nature of the juvenile firesetter problem; 2) the resources (e.g., manpower, money, space, etc.); and 3) the availability of private funding, if necessary. As noted earlier, no one program structure is best or even feasible for every jurisdictions. Large fire departments with the necessary people and funds can staff full-time juvenile firesetter programs and provide all of the necessary services using multiple personnel. Smaller departments or those with limited funds may be able to fund one full-time position or may have different fire service personnel assume some of the juvenile firesetter program responsibilities in addition to their other duties.

Like many juvenile firesetter program activities, recruiting representatives to serve on the coordinating council takes time. For some programs, especially smaller programs, it may take more time than the coordinator can supply. At a minimum, the program coordinator should contact each potential referral agency and describe the program. Referral networks need to be established if the juvenile firesetter program is going to meet the needs of the youth it sees. Each referral agency needs to understand that juvenile firesetting is a community problem and they must be willing to be part of the solution.

The juvenile firesetter program staff must also be prepared to handle the turf issues that may exist between agencies. These issues are often deeply-rooted and preclude agencies from working together. One of the major functions of the coordinating council is to maintain communication between agencies. Recruiting representatives from all of the key agencies listed above will help gain support for the program. All of the agencies must be involved in the planning and coordination stage of a juvenile firesetter program. This involvement will give each agency a vested interest in the success of the program and assist in breaking down the barriers that may arise over turf issues.

CHAPTER 3: SCREENING AND EVALUATION

Objectives

There are four major objectives to be achieved in the screening and evaluation of firesetting youth and their families. The first is the assessment of firesetting risk. A complete firesetting history must be taken to determine the extent and nature of the problem. In addition, a detailed description of the motives and circumstances surrounding the most recent firestart must be documented to ascertain the severity of the presenting problem. Based upon current information, an estimate must be made of the likelihood that firesetting behavior will recur.

The second objective is the evaluation of the psychosocial and environmental features related to firesetting behavior. Firestarting episodes do not happen as isolated incidents. Although the majority of juvenile firestarts are estimated to be the result of curiosity or accident, about one-third of juvenile fires are started by troubled and conflicted children. Therefore, for a selected proportion of firesetting youth, there must be an assessment of the underlying psychosocial features which accompany their firesetting behavior.

The third objective is the determination of criminal intent. If juveniles are involved in significant fires resulting in property loss, personal injury or death, then they are at risk for being arrested for the crime of arson. Several factors are taken into consideration for determining criminal intent, including whether firesetters have reached the age of accountability, the nature and extent of their firesetting histories, and the motive and intent of their firesetting. Although legal definitions of arson vary from state to state, if an evaluation reveals that there is sufficient evidence indicating malicious and willful firesetting, then the youth can be arrested for arson.

The final objective is the development of an intervention plan. The result of a comprehensive evaluation is the development of an effective intervention plan. Intervention plans must identify the specific steps to be taken to eliminate firesetting behavior and to remediate the accompanying psychosocial problems. In addition, adequate incentives must be set in place to insure that juvenile firesetters and their families will follow through with the recommended interventions.

The Target Populations

There are three general groups of juvenile firesetters which must be targeted for screening and evaluation. The first group is young children under seven years of age. The fires started by the majority of these children are the result of accidents or curiosity. In general, they do not exhibit significant psychological problems and their family and peer relationships are intact and stable. (There are a small number of children involved in firesetting who exhibit severe psychopathology, and these children are generally referred immediately for psychological evaluation and treatment.)

The second group of firesetters are children ranging in age from eight to twelve. Although the firestarting of some of these children is motivated by curiosity or experimentation, a greater proportion of their firesetting represents underlying psychosocial conflicts.

The third group of firesetters are adolescents between the ages of 13 and 18. These youth tend to have a long history of undetected fireplay and firestarting behavior. Their current firesetting episodes are either the result of psychosocial conflict and turmoil or intentional criminal behavior.

There are a number of different community agencies that screen and evaluate juvenile firesetters and their families. Based on their broader role in the network of community services, each of these agencies will have different functions regarding their work with firesetters. Consequently, the screening and evaluation methods they select will vary depending on their specific needs. We will concentrate here on the fire service and the screening and evaluation methods used by fire service personnel.

As noted earlier, the fire service is frequently viewed as the lead agency in the community for screening and evaluating juvenile firesetters. The primary role of the fire service is the early identification of firesetting youth and their families. There are a number of different ways in which juvenile firesetters are identified by the fire service. First, parents may discover firestarting behavior and voluntarily seek help for their children. Second, other community agencies may look to the fire service as the experts in working with juvenile firesetters and refer their cases to them. Third, fire and arson investigation efforts may reveal the involvement of juveniles in significant fires. Finally, if firesetters are arrested for arson, probation and juvenile justice may refer them to the fire service for an evaluation of the severity of their firesetting problem. Hence, the fire service is likely to see the entire range of juvenile firesetters from young children under seven who firestart out of curiosity to adolescents involved in recurrent firesetting.

Frequently the fire service will have two levels of screening and evaluation procedures. The first level involves preliminary screening to determine the immediate severity of the firesetting problem. This generally is done by telephone interview. If there appears to be no imminent risk, basic information is obtained and appointments are made for additional evaluation sessions. A complete interview with firesetters and their families follows and represents the second level of evaluation. The primary goals of these interviews are to analyze the severity of the firesetting behavior and describe the psychosocial environment. This two level system facilitates the handling of emergency problems and establishes the conditions necessary for a comprehensive evaluation system.

Fire Service Procedures

The following instruments and methods are used as screening and evaluation procedures by the fire service.

Telephone Contact Sheet

When parents or community agencies call the fire department to request help for firesetting youth and their families, this sheet is used as a preliminary screening mechanism. Basic information is gathered such as names, addresses, telephone numbers, a brief summary of the firesetting problem, and a description of the steps to be taken, which frequently includes the dates and times for setting-up personal interviews or the other follow-up procedures which are to be implemented for particular cases.

USFA 's Interview Schedules

These interview schedules are designed to provide the juvenile firesetter program with systematic methods for evaluating juvenile firesetters and their families. The interview schedules consist of a series of questions which are asked of tiresetting youth and their families in personal interviews, The application of these interview schedules yields information regarding the severity of the firesetting problem and preliminary data on the psychosocial environment of juvenile firesetters and their families. The USFA interview schedules have been widely used by a number of fire departments throughout the country and represent standard practice for many tire departments and juvenile firesetter programs. With minimal training, these procedures can be used by fire service personnel to screen, evaluate and refer juvenile firesetters and their families to appropriate service agencies in the community. There are three manuals describing in detail the application of these interview schedules. These manuals can be obtained by calling or writing the U.S. Fire Administration in Emmitsburg, Maryland.

These interview schedules and the manuals which describe their application are divided into three age groups. The first manual outlines the interview schedules and methods for working with children seven and under. The second manual describes the interview schedules and methods to be applied to children ages seven to thirteen. The third manual contains the interview schedules and procedures for working with adolescent firesetters. The implementation of the procedures in each of these three manuals allows fire departments to screen and evaluate the entire range of juvenile firesetters.

These interview schedules and accompanying manuals are used by fire departments extensively throughout the United States. Many departments have established juvenile firesetter programs following the guidelines suggested in these manuals. The interview schedules also contain information which may be useful to mental health professionals regarding the severity of the firesetting problem and the conditions of the psychosocial environment. Juvenile firesetters and their families are interviewed alone and together for approximately ninety minutes by fire service personnel using the interview schedules. The interview schedules are organized and presented in slightly different ways depending on the age of the firesetter. For children under seven, the interview schedule is divided into two sections. Section one focuses on questions regarding firesetting behavior and section two requires observations to be made regarding the home and the parents. For children seven through thirteen, the interview schedule is divided into three sections. The first section asks questions related to firesetting history, the second section presents questions related to the home and family, and the third section asks questions regarding school and peers. For adolescent firesetters the interview schedule is divided into two main sections. Section one asks questions related to firesetting history and details of the most recent firesetting incident. Section two asks questions regarding the psychological environment, including information about physical health, the home, the family, peers, and school. For all age levels, parents are asked to complete a questionnaire which contains observations about the psychological behavior of their children.

The interview schedules have scoring procedures which classify firesetting youth and their families according to risk levels. These risk levels refer to the probability that the juvenile firesetters are likely to participate in future firesetting incidents. There are three levels of risk -- little, definite, and extreme -- representing increasingly severe firesetting behavior. In general, children classified as little risk firestart by accident or out of curiosity, and require educational intervention to remediate their problem. Juveniles classified as definite and extreme risk firestart because of psychological conflict, family difficulties or as part of a pattern of anti-social and delinquent behavior. They require mental health or juvenile justice intervention. Hence, the interview schedules yield a specific method for classifying juvenile firesetters according to the severity of their presenting problem. This classification system also suggests the type of interventions most likely to be beneficial to tiresetting youth and their families.

The primary advantage of the interview schedules is that they provide systematic procedures for fire service personnel to evaluate the entire range ofjuvenile tiresetters and their families. In addition, these interview schedules yield a quantifiable method for classifying the severity of the firesetting problem and for recommending specific types of interventions. Also, only a brief training period is required to teach fire service personnel how to use these interview schedules. Their application is well documented in three manuals and they are widely accepted and applied throughout the fire service.

The major disadvantage of these interview schedules is that their validity and reliability have not been investigated. Therefore, the accuracy and consistency of the information which they yield remains open to question. A primary concern is that children identified as little risk may actually be exhibiting signs of more serious firesetting behavior. One way to address this problem is to monitor little risk firesetters for a period of time subsequent to their evaluation and educational intervention. Also, in cases where definite and extreme tiresetting youth have been identified, it is recommended that other assessment strategies be used in conjunction with the interview schedules. For example, as a general rule these types of cases should be referred for additional evaluation by mental health professionals. While these interview schedules provide an important first step in screening and evaluating juvenile firesetters, virtually nothing is known about the quality of the information they yield. Therefore, back-up procedures must be set in place. In many cases, this includes being able to consult with a mental health practitioner for cases which need additional clarification. This will insure that juvenile firesetters and their families receive appropriate assessment and intervention.

CHAPTER 4: INTERVENTION SERVICES

Purpose

The Intervention Services component presents the primary intervention strategies designed to reduce the incidence of juvenile involvement in firesetting behavior and arson-related activities. These strategies reflect three major intervention approaches. The first strategy is primary prevention. The goal of primary prevention is to provide substantial fire safety and educational experience to juveniles so that they develop fire-competent behaviors and avoid participation in unsupervised firestarts. The second strategy is early intervention. Youth participating in fireplay and firesetting behavior motivated by accident, curiosity or experimentation can be identified and educated to reduce the likelihood of their future involvement in unsupervised firestarts. The third strategy is core intervention. Recurrent firesetters frequently experience significant psychological and social conflict and turmoil related to their firestarting activities. It is hypothesized that if these psychosocial problems can be adjusted or remediated, then not only are the chances of involvement in future firesetting episodes greatly reduced, but the quality of life is likely to improve for these juveniles and their families. ISA recommends that such core intervention be delivered by mental health or other social service providers.

The three intervention strategies - primary prevention, early intervention, and core intervention - are aimed at reducing juvenile involvement in the entire range of unsupervised fireplay and firesetting activities. Each of the strategies have specific intervention objectives and they are aimed at particular target populations. Primary prevention efforts involve several community agencies including the schools, the fire service, and law enforcement. The fire service is the lead community agency providing early intervention services. They use many different types of program models to work with juvenile firesetters. Core intervention services involve mental agencies and professionals and the probation and juvenile justice systems.

Each of the three intervention strategies are designed to achieve specific objectives. Primary prevention efforts are intended to reduce the incidence of first-time unsupervised fireplay and firesetting in populations of otherwise normal youth. This is accomplished by providing children of all ages with educational experiences focused on the rules of fire safety and prevention and understanding the consequences of fireplay and firesetting.

Early intervention programs are focused on identifying both children at-risk for fireplay or firesetting activities and those involved in first-time fireplay and firesetting episodes. In addition, their objective is to prevent the recurrence of fireplay and firesetting incidents. The implementation of short-term evaluation, education, and referral mechanisms within the fire service and other supporting community agencies are designed to meet these objectives.

Core intervention services are aimed at eliminating recurrent firesetting behavior and providing treatment and remediation for the contributing psychosocial determinants. Mental health intervention is the primary method utilized to stop recurrent firesetting and treat the underlying causes of the behavior. Probation and juvenile justice efforts provide legal incentives to youth and their families to pursue treatment for their patterns of antisocial and delinquent behavior. If treatment recommendations are not followed, the juvenile justice system can implement legal consequences and punishments related to firesetting and arson offenses.

Situational Influences

The design and implementation of juvenile firesetter programs will depend upon the commitment of time and resources participating agencies are willing to make in their community. For example, schools must decide whether primary prevention programs designed to teach fire safety are a high priority for their curriculum. Fire departments, heavily committed to suppression activities, will need to direct their focus to the prevention aspect of fighting fires. Law enforcement, probation and juvenile justice must select to pay particular attention to the firesetting population of juveniles, as opposed to other groups of delinquent youth. Frequently additional program efforts aimed at specific problems areas or target populations can be incorporated into existing operations, thereby keeping costs at a minimum. Nevertheless, the level of time and resources committed to juvenile firesetter intervention programs is directly related to their content, utility, and effectiveness.

Juvenile firesetting must be viewed as a community problem, and as such, it deserves community-wide attention. Although fire departments may take the lead role in developing programs for juvenile firesetters, their efforts alone will not resolve the problem. It is crucial that there be working linkage established between the various community agencies capable of helping juvenile firesetters and their families. Schools, the fire service, law enforcement, juvenile justice and mental health must all establish open communication channels with one another so that an organized effort is mounted to reduce juvenile involvement in firesetting and arson-related activities.

Critical Issues

The success of juvenile firesetter intervention programs depends on several factors. First, the community must be educated about the problems of juvenile firesetting. An effective public relations campaign must be developed to teach parents and adults how to recognize the problem in children and where and how to go for help to resolve it. Regardless of the level of intervention, from primary prevention to core intervention, the public must understand the seriousness of juvenile involvement in firesetting and they must be knowledgeable enough to take the first steps to get the appropriate help.

Each community agency focusing their attention on the problem of juvenile firesetting is likely to have slightly different roles and responsibilities, depending on the nature and extent of their services. Those agencies and professionals involved in helping juvenile firesetters must be trained in how to work with this special population of youth. Although training needs will vary according to the type of services offered, designing and implementing intervention programs often requires special expertise and information. Education manuals coupled with training seminars are important resources for establishing and maintaining effective intervention services.

Primary Prevention

Primary prevention programs are aimed at reducing juvenile involvement in first-time unsupervised fireplay and firesetting incidents. The basic premise of these programs is that if children understand the rules of fire safety and prevention and the consequences of firestarting, they are less likely to initiate or participate in nonproductive firesetting. Primary prevention efforts are educational programs designed to teach children of all ages fire safety and survival skills.

There are several different educational models utilized in primary prevention programs. The models employed largely depend on the sites which operate the programs. Primary prevention programs are found in the schools, the fire service, and law enforcement. Schools can offer a wide range of prevention activities including fire safety education curriculum and activities, slide presentations, films, and assemblies. The fire service can mount national and local media campaigns, utilize district fire houses to provide tours and educational seminars for youth, and work with their school districts to present unique educational experiences. Law enforcement can incorporate fire safety education as part of their general anti-crime efforts aimed at youth. Primary prevention programs can utilize a variety of different learning strategies and activities to accomplish the common objective of teaching youth how to develop fire-safe and competent behaviors.

It is recommended that community organizations or agencies launch a comprehensive fire prevention effort designed to reach a broad age-range of children. Educational programs for preschool children should be explored as well as programs aimed at elementary, middle and high school aged youth. Schools are the obvious site where maximal efforts can be focused to reach the majority of children. The amount of time set aside for teaching fire prevention and safety will depend on the level of effort schools are willing to commit. A minimal effort might consist of a fire education presentation to youth coupled with the distribution of printed material to parents. A more comprehensive approach might be the adoption of a fire safety curriculum. There are several excellent packages of fire safety and prevention programs already developed for schools. The particular program or set of programs developed depends on the available resources and the range and depth of desired services. It is strongly recommended that schools integrate primary prevention efforts into their on-going curriculum plans.

Fire service efforts can be important adjuncts in helping to promote the development of fire safety behaviors in children and their families. For example, parents who first notice their children's interest in fire or who have found their children playing with matches may instinctively call their local fire department for help. Fire departments can offer to talk with these youth, have them tour the local fire house and provide short-term educational services designed to teach tire prevention to children and their families. In addition, the fire service can work with their local schools to enrich fire prevention programs by offering classroom visits or assemblies, slide presentations and films designed to communicate information on fire safety and prevention. Finally, local fire departments can support national programs, such as National Fire Prevention Week, by mounting active print and television media campaigns designed to promote fire safety.

There are a wide variety of primary prevention programs currently operating in communities throughout the country. A sample of these programs include: 1) Learn Not to Burn, developed by the National Fire Prevention Association; 2) Knowing About Fire, developed by the National Fire Service Support System; 3) Fire Safety Skills Curriculum, used by the Oregon State Fire Marshal's Office; 4) Juvenile Crime Prevention Curriculum, developed in St. Paul, Minnesota; and 5) Kid's Safe Program, used by the Oklahoma City Fire Department. These and other programs are listed in the Resource List at the end of this chapter. Programs are encouraged to contact the U.S. Fire Administration and juvenile firesetter programs around the country.

Early Intervention

The fire service is the leading community agency involved in the development of early intervention programs for juvenile firesetters. The primary objective of early intervention programs is to identify children at-risk for participating in unsupervised fireplay and firesetting incidents. In addition, these programs are aimed at preventing the recurrence of first-time firesetting episodes motivated by accident, curiosity or experimentation. These objectives are accomplished by setting-up short-term evaluation, education, counseling and referral services designed to stop firesetting behavior and identify related psychosocial problems.

As noted earlier, ISA advocates an evaluation, education, and referral program model. This approach is the one most frequently employed by fire departments across the country and is the recommended strategy for building effective juvenile firesetting programs. To implement this model, fire departments must establish methods for screening and evaluating the firesetting risk of children and their families. The recommended screening methods are discussed in the Screening and Evaluation chapter. The identification of risk levels allows fire departments to determine the most appropriate strategies for remediating the current firesetting problem. If children are identified as little risk, then it is likely that short-term education intervention will stop any further firesetting behavior. Fire departments have successfully implemented a number of different educational programs. If youth are identified as definite or extreme risk, while they may benefit from educational programs, they are likely to need core intervention services. Fire departments must know how to refer firesetters and their families to the appropriate service agencies.

Again there are many programs that have implemented evaluation, education, and referral programs. The following programs are examples of evaluation, education and referral interventions for juvenile firesetters which have operated in fire departments across the country. They were selected because certain features of these programs represent outstanding or exceptional aspects of the evaluation, education and referral program model. Additional information can be obtained by contacting the programs listed in the Resource List at the end of this component.

A. The Juvenile Firesetter Program, Columbus, Ohio

The primary purpose of this Juvenile Firesetter Program is to prevent juveniles who are setting fires and playing with matches and lighters from staring additional fires. The majority of these juveniles are referred from fire investigators, with a smaller number coming from children's services and mental health agencies. These children and their families are evaluated using USFA's Interview Schedules. All youth attend 4-6 educational sessions. This educational segment of the program is one of the outstanding features of its operation. Those youth identified as definite or extreme risk are referred for further core intervention services. Follow-up evaluation forms are sent every six months for two years to participating families and the resulting data indicate a 7% recidivism rate.

While the format of this juvenile firesetter program represents a standard example of the evaluation, education and referral program model, the educational feature of this approach deserves special mention. Prior to their participation in the educational sessions, juveniles complete written pretests designed to assess fire safety knowledge. They then attend four to six educational sessions at their local firehouse, depending on their age and the history of their firesetting behavior. Audio-visual teaching aids are used extensively. *The Official Fire Safety Manual*, containing games and puzzles designed to teach fire safety and prevention rules, is used with all youth. Separate manuals have been developed for children 4-6 years, 7-9 years, and 9-12 years. In addition, the family does homework, including designing a Home Fire Escape Plan and conducting a home fire safety inspection. In the final educational meeting, children complete post-tests to assess the amount of increased knowledge of fire safety and prevention accrued from the program. This represents a comprehensive approach by fire service personnel to provide educational experiences for children at the firehouse.

B. Operation Extinguish, Montgomery County, Maryland

Operation Extinguish is one of the programs run by the Montgomery County Fire Department's Division of Fire Prevention. Youth are referred to fire prevention by their parents, from the youth division of the police department, and from the juvenile services administration. All juveniles are evaluated using family assessment methods and following the guidelines recommended by USFA's Interview Schedules. All children also attend fire safety classes conducted by the division of fire prevention. These classes represent an outstanding feature of this program. Families of juvenile firesetters are referred to a private mental health agency for at least six family counseling sessions. This also is a unique program feature. Firesetters and their families are referred for other services on an as-needed basis. Release from Operation Extinguish is contingent upon completion of the prescribed intervention plan.

Operation Extinguish has two unique program features. The first is a highly structured format of fire education classes. Three two-hour fire safety classes are run for groups of juvenile firesetters. Firesetters attend the first two classes and may bring their siblings. Parents also attend the last class. Audio-visual aids are used along with a manual, *A Question of Burning.* Week one covers the history of fire facts. Fire prevention, recognition of fire hazards, escape planning and survival techniques are discussed. Arson and arson laws in the state of Maryland are reviewed. The homework, to be completed by week three, is assigned and consists of developing a home fire escape plan. Week two focuses on burn injuries. Children participate in writing exercises designed to help them think about the potential consequences of firesetting. Week three summarizes fire safety rules for parents. Escape plans are reviewed and the importance of knowing how to react in fire emergencies is discussed. By the end of the third session, both children and parents report being satisfied with the educational experience.

The second unique program feature is that the majority of firesetters and their families are referred for at least six family counseling sessions. The entire family is encouraged to participate in these sessions, since it is likely that siblings also may be involved in fireplay and firesetting activities. Counseling sessions are tailored to meet the individual needs of families. Families participating in these sessions report that their communication is greatly improved as a result of these counseling sessions.

C. Fire-Related Youth (FRY) Program, Rochester, New York

The FRY program is housed in the Rochester Fire Department. It receives the majority of its referrals from within the fire department. When the program receives referrals, fire investigators conduct complete investigations of the firesetting incidents. In addition to investigating the scene and conducting a records check, investigators interview children and their parents. The interviews are not designed to draw definite conclusions about the psychosocial functioning of juveniles, rather they are intended to provide investigators with more information about the fire. The USFA Interview Schedules are used as guidelines during these interviews. After interviewing the parents, investigators meet with the child to talk about the incident and provide fire safety education. The exact nature of the education depends on the age of the child. The majority of juvenile firesetters interviewed are referred for additional services.

There are two outstanding features of the FRY program. The first is their well-established linkages with numerous service agencies within their community. The FRY program can refer juvenile tiresetters to one of four mental health agencies, the Police Department's Family and Crisis Intervention Team, Child Protective Services, Probation, or Family Court. This referral system also includes a dual waiver form that allows a free flow of information between the FRY program and all referral agencies. The solid referral network established by the FRY program ensures that children and their families will receive the services necessary to stop firesetting and remediate the related psychosocial problems.

The second exceptional program feature is the complete documentation of the FRY program in two sequential manuals. These manuals not only describe the operation of the program, but they report investigative studies concerning a number of different topics including a complete description of the population of firesetting juveniles and explanations of their firestarting behavior. These manuals provide visibility and credibility for the FRY program.

D. Operation Fire S.A.F.E., Oklahoma City, Oklahoma

Operation Fire S.A.F.E. was one of the three implementation sites selected to participate in the National Juvenile Firesetter/Arson Control and Prevention Program National Evaluation. The program is targeting the Central Oklahoma area, which includes four counties (Oklahoma, Canadian, Cleveland, and Logan) and thirty-five fire departments. The program has established an Operational Committee comprised of the area fire departments and the Association of Central Oklahoma Governments (ACOG). This committee has primary responsibility for developing the program. Individual members or subcommittees have taken responsibility for developing specific aspects of the program, including video development, site selection, logo development/selection, development of the initial contact and evaluation forms, and working with the mental health guidance centers and the District Attorney's Office.

Operation Fire S.A.F.E. is targeting a very large, diverse area with multiple fire service, mental health, juvenile justice, human service, and police agencies. Operation Fire S.A.F.E. established twelve geographically diverse sites to provide assessment and fire safety education to the area's juveniles. Fire departments in the central Oklahoma area that are not equipped to handle juvenile tiresetters can refer them to one of the seven sites. Each site will have 1) a set of U.S.F.A. screening instruments; 2) a video library; 3) TV/VCR; 4) standardized forms developed by Operation Fire S.A.F.E. (initial contact, interview form, information release waiver, summary form); and 5) toys.

Operation Fire S.A.F.E. is fortunate to have the support of the area's mental health professionals. The program has been working with the county guidance centers to develop a consistent regional system for referring families in need of mental health counseling. The Guidance Center Coordinator developed a six-month pilot referral program for the Oklahoma City Fire Department, which is being implemented. That department refers families to one guidance center for 12-week group therapy. The remaining fire departments are continuing to refer families to the other area guidance centers, clinics, and individual practitioners. After the pilot program is evaluated, it may be picked up by the other guidance centers.

In addition, the Oklahoma County Guidance Center completed a five-part training series for the Oklahoma City Fire Department Fire S.A.F.E. interviewers on effective interviewing techniques. The guidance center counselors have offered the same course to other area departments involved in Operation Fire S.A.F.E. The guidance center will also be offering a monthly discussion group for Fire S.A.F.E. interviewers to discuss their feelings and frustrations.

The program has also been working with the Oklahoma County District Attorney's Office. One area of concern was Operation Fire S.A.F.E.'s policy statement which asserted that any information collected as part of the Operation Fire S.A.F.E. assessment be kept strictly confidential and not be used for fire investigation. Although the statement has been accepted, the program has been cautioned that the fire investigators should complete their investigations before referrals to the program are made. Operation Fire S.A.F.E. is also trying to establish a mechanism that would provide "motivation" for parents who refuse or are reluctant to have their child participate in the program after a fire has been set. The program is working with the Department of Human Services to encourage involvement. When parents are extremely reluctant, the case is treated as a "neglect" case by DHS.

One key component of the juvenile firesetter program is the development of a video describing Operation Fire S.A.F.E. The eleven minute video, developed as part of this program, will provide information to the media, school administrators and counselors, parents, mental health professionals, and other agencies that work with juvenile firesetters on the purpose of the program, how to access the program, the locations of assessment/education sites, etc. Operation Fire S.A.F.E. also developed a program brochure and poster to accompany the video which will be used as part of the program's public education campaign.

Core Intervention

Children and adolescents involved in recurrent firesetting behavior and displaying serious psychopathology are candidates for core intervention services. In addition, youth whose firesetting is willful or malicious and an expression of criminal intent also are likely participants for core intervention. Core intervention services are those modalities which provide long-term help for juvenile firesetters and their families to eliminate firesetting behavior and remediate the accompanying psychopathology. There are two major modalities of core intervention, mental health treatment and the probation and juvenile justice system. If the juvenile firesetter program believes that a child they have screened is in need of core intervention services, that child should be referred to the appropriate agencies. Establishing referral mechanisms is covered in the next chapter.

Other Considerations

When developing an early intervention program, juvenile firesetter programs will have a vast amount of educational materials from which to choose those most appropriate for their service needs. Program staff should consider contacting the U.S. Fire Administration, the National Fire Protection Agency, and other fire departments to inquire about available materials. Although ISA does not advocate one set of materials over another, we do feel strongly that certain approaches should be avoided. We do not believe that scare tactics or the use of graphic pictures of burn victims are an effective way to teach fire safety education. Although some programs do use these techniques, we do not feel they are appropriate, especially for young children. In general, programs have move away from having children visit burn victims in local bum units. Such visits are not seen as fair to the bum victim or as an effective educational tool for the firesetter. On a different note, programs also sometimes have children tour the local fire station and sit on the fire truck. Children usually enjoy these tours and they can be viewed as a reward. Tours should only be given AFTER the child has completed their fire safety education. Children should not be rewarded for setting a fire, but could participate in a fire station tour at the completion of their intervention.

One final factor to consider in developing intervention services is insuring follow-through by firesetters and their families. At the very least, agencies referring juveniles and families for additional intervention can follow through on their own with telephone calls to the referral agency to ascertain whether the recommended contact actually occurred. Different community agencies may have various ways of providing incentives or creating leverages so that follow-through with the intervention plan is more likely to occur. For example, fire departments can offer children a tour of the firehouse or a visit with the fire chief if they pursue their educational intervention. Law enforcement, probation, and juvenile justice can help youth avoid arrest and incarceration by diverting them to mental health counseling. Mental health can suggest to firesetting youth and their families that their quality of life is likely to improve as a result of participation in treatment. Schools can refuse to accept firesetters in their classrooms. There are a number of leverages which can be successfully implemented in certain circumstances to help insure follow-through with the intervention plan.

Resource List

Primary Prevention

School Curriculum and Programs

1. *CTW'S Fire Safety Project*
   Sesame Street Fire Safety Resource Book

   Contact: Children's Television Workshop
   1 Lincoln Plaza
   New York, NY 10023
   (212) 595-3456

2. *Learn Not to Burn*

   Contact: National Fire Protection Association
   1 Batterymarch Park, P.O. Box 9101
   Quincy, MA 02269
   (617) 770-3000

3. *Knowing About Fire*

   Contact: Paul Schwartzman
   National Fire Service Support Systems, Inc.
   20 North Main Street
   Pittsford, NY 14534
   (716) 264-0840

4. *Fire Safety Skills Curriculum*

    Contact: Judy Okulitch
    Program Manager
    Office of the State Fire Marshal
    3000 Market Street, NE, #534
    Salem, OR 97310
    (503) 378-3475

5. *The Juvenile Crime Prevention Curriculum*

    Contact: Public Relations Department
    The St. Paul Companies
    385 Washington Street
    St. Paul, MN 55102

6. *Follow the Footsteps to Fire Safety*

    Contact: City of St, Paul
    Department of Fire and Safety Services
    Fire Prevention Division
    100 East Eleventh Street
    St. Paul, MN 55101
    (612) 228-6203

7. *Project Open House*

    Contact: Richard A. Marinucci
    Farmington Hills Fire Department
    28711 Drake Road
    Farmington Hills, MI 48331-2525
    (313) 553-0740

8. *Kid 's Safe Program*

    Contact: Fire Safety Education Curriculum for
    Preschool Children
    Oklahoma City Fire Department
    Public Education
    820 N.W. 5th
    Oklahoma City, OK 73106
    (405) 297-3314

Fire Service Programs

1. *National Fire Prevention Week*

2. *Curious Kids Set Fires*

    Contact: US Fire Administration
                 National Fire Academy
                 16825 South Seton Avenue
                 Emrnitsburg, MD 21727

3. *Big Fires Start Small*

    Contact: National Fire Protection Association
                 1 Batterymarch Park, P.O. Box 9101
                 Quincy, MA 02269
                 (617) 770-3000

4. *Firebusters*

    Contact: Earl Diment
                 Office of Community Relations
                 Portland Fire Bureau
                 55 Southwest Ash
                 Portland, OR 97204
                 (503) 823-3700

5. *Public Fire Education Today*

    Contact: US Fire Administration
                 National Fire Academy
                 16825 South Seton Avenue
                 Emmitsburg, MD 21727

Fire Department Programs in Schools

1. *Slide Presentations*

    Contact: Office of the Fire Chief
                 Fourth Floor East
                 Largo Government Center
                 9201 Basil Court
                 Landover, MD 20785

2. *Films*

    Contact: Juvenile Firesetter Program
                 Fire Prevention Division
                 Fire Marshal's Office
                 301 2nd Avenue South
                 Seattle, WA 98104
                 (206) 296-6670

3.   *Assemblies*

    Contact: Juvenile Firesetter Program
                         Fire Prevention Division
                         301 2nd Avenue South
                         Fire Marshal's Office
                         Seattle, WA 98104
                         (206) 296-6670

    Contact: Captain Henry Begroot
                         Fire Prevention
                         San Jose Fire Department
                         4 North 2nd Street, Suite 1100
                         San Jose, CA 95113
                         (408) 277-4444

Law-Enforcement Programs

1.   McGruff

    Contact: The National Crime Prevention Council
                         1700 K Street, N.W., Second Floor
                         Washington, DC 20006

Early Intervention

Evaluation Education and Referral Programs

1.   *The Juvenile Firesetter Program, Columbia, Ohio*

    Contact: Lonnie Poindexter
                         Juvenile Firesetter Program
                         Bureau of Fire Prevention
                         300 N. Fourth Street
                         Columbia, OH 42315
                         (614) 645-7641

2.   *Operation Extinguish, Montgomery County, Maryland*

    Contact: Mary Marchone
                         Division of Fire Prevention
                         101 Monroe Street
                         Rockville, MD 20850
                         (301) 271-2442

3.   *Fire Related Youth (FRY) Program, Rochester, New York*

    Contact: Jerold Bills
                         FRY Program
                         Rochester Fire Department
                         Room 365
                         Public Safety Building
                         Civic Center Plaza
                         Rochester, NY 14614
                         (716) 428-7103

4. Juvenile Firesetter Program, Portland Oregon

   Contact: Don Porth
   Portland Fire Bureau
   55 Southwest Ash
   Portland, OR 97204
   (503) 823-3700

5. Operation Fire S.A.F.E.

   Contact: Jane Sutter
   Association of Central Oklahoma Governments
   6600 North Harvey Place, Suite 200
   Oklahoma City, OK 73116
   (405) 848-8961

CHAPTER 4: REFERRAL MECHANISMS

Introduction

Juvenile Firesetter Programs should occupy a central position between the *sources* of juvenile firesetters (fire service, schools, parents) -- the people who detect the firesetter -- and the *target agencies* (counseling services, juvenile court, etc.) -- the agencies or people who provide specialized treatment or sanctions to the juvenile firesetter.

In most programs a substantial number of firesetters will be detected by fire service personnel and brought to the program for some education/intervention; consequently, many of the "referrals" occur within the fire service and do not require the assistance of others, either in finding the firesetters or addressing their problems. The typical case of this kind is the young child without any significant pathology who is identified by a fire investigator, referred to the program, screened by the fire service, provided fire safety education, and released. In many jurisdictions these may be the most frequent kind of cases. However, all other cases require effective referral systems so that (a) people outside the fire service will bring the firesetter to the attention of the program, and (b) the juvenile firesetter who exhibits serious problems of adjustment or delinquency can receive the appropriate additional resources. The screening and intervention activities conducted within the program represent the very heart of any program, but they are not at all sufficient to insure that all firesetting youths are receiving the help (or sanctions) they deserve. Indeed, without a wide range of referral sources the program will never see a sizable segment of the juvenile firesetter population, and without the appropriate agencies and individuals to whom youths can be sent for additional help, many firesetters (especially the more troubled youths) will never receive the services they need.

A graphic depiction of the desired referral network is shown in Figure 4.1. Typical sources of referrals are shown on the left, with the fire service typically providing most of the referrals, followed by parents, schools, etc. On the right are the major types of referral targets, agencies that can provide the appropriate additional services to the juvenile firesetter and the family.

Several points of this figure deserve special attention. First, it shows a large number and variety of possible referral sources and target agencies. Although these may vary widely across jurisdictions, it is important that the program give serious consideration to developing referral arrangements with all such agencies and groups. Second, two of the three key target agencies -- social services and mental health agencies -- may also serve as referral sources as well. Third, this depiction is necessarily an over simplification of the actual sequence of events involved in many referrals. For example, as shall be discussed below, the arrangements among the criminal justice agencies, the program, and the mental health and social service agencies can be quite complicated, involving several decision points and transfers for any one case. Fourth, some referral mechanism will be for the purpose of Fig. 4.1 here merely recording and tracking juvenile firesetter cases that are not actually seen by the program. Although we feel that the juvenile firesetter program should be able to screen all firesetters, often they do not. At a minimum, ISA feels that the juvenile firesetter program should be aware of every fire set by a juvenile in its jurisdiction.

Identifying Referral Sources and Target Agencies

The foundation for the referral mechanisms is laid in the early stages of program planning and development. The first step in the development of effective referral mechanisms is the identification of all potential referral sources and target agencies. The types of agencies and groups shown in Figure 4.1 should serve as a start for the identification of referral sources and agencies to be developed. Once a list of referral types is developed, the program should complete a worksheet providing the names of specific organizations, a description of the preferred referral arrangement, the individuals who will serve as the primary contact or liaison to the program, and their address and telephone number.

To some degree, much of the identification process will have been accomplished during the planning and coordination stage. However, the individuals from the various agencies who will serve as the functional contacts for referrals may be different from those who are first involved in planning and coordination and the actual approval of the referral relationship. Therefore, it is likely that at least two levels of agency officials will be involved in the development of these mechanisms: (1) Relatively high-level officials with the authority to bind the agency to a referral agreement, and (2) individuals who will have continuing responsibility for the operation of the referral mechanism.

Although the selection of referral agencies will depend on the particular nature of the juvenile fire setter problem in the area, virtually all programs in jurisdictions of medium-to-large size should explore the possibility of referral arrangements with all the types of agencies listed in Figure 4.1. The only exceptions will be small jurisdictions that do not have social and mental health service agencies or the multiple levels of the criminal justice system. If counseling and therapy services are not available locally, the program should investigate the possibility of referral arrangements with such agencies in neighboring towns and cities.

The target agencies selected will also be dependent, at least to some degree, on the nature and severity of the problems of the juvenile firesetters; e.g., the extent to which the youths display serious problems of adjustment and delinquency, whether family counseling is required, etc. We suggest, however, that the referral mechanism be arranged (at least the groundwork laid) for all the major target agencies so that if the need arises, the resources will be there.

Target referral agencies -- places to which the firesetters are to be sent for special services -- should be screened carefully for quality. If they are government operated (e.g., a community mental health center) there is less concern about quality, since these organizations typically have to meet standards that are carefully developed and regularly monitored. Private social service and mental health organizations (which are increasing in number) and individual practitioners should be carefully screened before sending them referrals. Practitioners should be certified and/or licensed in their respective fields -- social worker, psychologist, etc. If the organizations have referral relationships with other programs or institutions in the area, you can call them and ask for their opinion of the quality of services.

You should also examine the particular capabilities and capacities of the target referral agencies -- are they equipped to handle the cases you may be sending them?

Contacting Referral Agencies

The source and target agencies should first be contacted by telephone and mail to present the basic idea to them. The emphasis should be placed on the special needs of the firesetter population and the mutual benefits of a referral arrangement for the program, the agency, the firesetter and his/her family, and the community at large. Face-to-face meetings should then be arranged to discuss the desired referral arrangement. The central purpose of these meetings is twofold: (1) To convince the agency of the importance of a referral arrangement; and (2) to communicate the nature of the referral relationship. For source referrals, it is important to provide detailed guidance on the characteristics of the youths to be sent to the program and the circumstances under which they are to be referred. For target agencies, it is important to describe the types of youth that will sent to them and the services they are likely to need.

Some juvenile firesetter programs have gone to extreme lengths to convince agencies to send referrals to the program. In Rochester, the program told agencies that if they did not agree to send referrals to the program, "the next death caused by a juvenile firesetter would be on their conscience.' Although this tactic is not appropriate for all circumstances, programs should adopt a programmatic approach, doing whatever works (within ethical boundaries). Another approach is to make use of the personal and political connections that the program has with influential officials at the agencies. For example, the Fire Chief can be very influential in securing the cooperation of other public safety officials. Once the program has gained a toehold in the mental health or educational community, supportive members of those communities can be helpful in recruiting other individuals and agencies into the referral network.

Developing Detailed Referral Agreements

After the agencies have agreed in principle to the referral arrangement and the details of the arrangement have been discussed, a written agreement should then be drawn up that specifies clearly the nature of the relationship and the specific responsibilities of each party. These agreements need not be elaborate legalistic documents; in most situations a single-page agreement will suffice. In some instances, initial agreements may be unwritten, oral agreements, but written statements of understanding should be developed at some point.

At some point in the discussions, the liability issue should be discussed in some detail with the agency. Written waivers of liability may be appropriate in certain cases. If in doubt about the proper course of action with respect to liability, you may want to consult a knowledgeable attorney.

The establishment of a referral agreement is only the beginning of a referral relationship between the program and the agency. The arrangement will be effective only so long as it is cultivated and maintained through continuing contact with agency officials. In particular, it is important to provide timely and informative feedback to the source referral agencies about the status of youths referred to the program -- results of screening, intervention outcomes, referrals out to other agencies, etc. Periodic meetings with the representatives of all agencies who are part of the juvenile firesetter referral network will also help to maintain the relationships, and will also provide a vehicle for addressing any problems before they become serious. Cases conferences are one possible strategy for maintaining communication and strengthening the referral network.

Some juvenile firesetter program have parents sign waivers allowing the referral agencies (both referral source and target referral agencies) to share information with the juvenile firesetter program. These waivers or releases permit the juvenile firesetter program to inform the referral source that the youth was assessed by the program staff and allows the program to forward the results of their assessment to a treatment agency. In addition these releases allow target agencies, such as mental health facilities and child protective services, to apprise the juvenile firesetter program of the status of a case. This exchange of information will enable program staff to monitor each case and ensure that referral linkages are successfully accomplished and no youth falls between the cracks. Problems have developed in some jurisdictions when youth are referred to the juvenile firesetter program and the referral source is never informed about the outcome of the case.

Differences Across Types of Jurisdictions

As with virtually any facet of the juvenile firesetter program, the nature and extent of referral mechanisms will be dependent upon the characteristics of the community in which it operates. Key characteristics influencing the referral mechanisms include: (a) the nature and severity of the juvenile arson problem, (b) the size of the jurisdiction, and (c) the availability of relevant resources.

The central factor in this regard is probably the size of the jurisdiction, which may range from small towns in rural areas to major metropolitan areas. The discussion above is most relevant to the medium-to-large cities where most of the juvenile firesetting is concentrated. In small, rural towns the problem of juvenile firesetting is likely to be less severe than in the larger cities -- both the incidence of firesetting and the seriousness of the youth's problem -- so a huge network of complex referral networks will probably be neither needed nor available. The major types of referrals as shown in Figure 4.1 are applicable to small towns as well, although the sheer number of agencies and individuals will be considerably fewer than in the larger cities. Consequently, the work of developing and maintaining the referral network is likely to be less difficult and time-consuming. On the other hand, many of the target agencies where the youths are sent for special services may be located in other towns and cities. Identifying these agencies and working out practical referral arrangements with them may require considerable time and effort tracing down the best and most appropriate resources. With respect to the counseling and therapy resources, the program may consider identifying one or two individual therapists (rather than entire agencies) who could provide most of the services.

CHAPTER 5: PUBLICITY AND OUTREACH

Purpose

This chapter describes how juvenile firesetter programs can develop a public information and education campaign to raise the public awareness about juvenile firesetting and the juvenile firesetter program. Surprisingly, many communities are often unaware of the juvenile firesetter problem or misinformed about the characteristics of firesetters. Parents may be reluctant to obtain help for their children suspected of firesetting for fear that they will be "put away." What many parents do not realize is that the majority of fires set by children are set out of curiosity. Without proper identification and education, however, simple curiosity can have deadly consequences.

The juvenile firesetter program has a responsibility to the community to inform them that a program exists to help juvenile firesetters. It is important for the community to understand that juvenile firesetter programs are designed to provide education for young firesetters and identify and refer troubled firesetters to counseling if necessary. Many juvenile firesetter programs are hindered because the community is unaware of the services they provide. This component will outline strategies that can be used to inform and educate the public about the program and the services it provides.

A note of caution --juvenile firesetter programs must be fully prepared to handle the requests for information and referrals generated from a publicity campaign. Programs must take care not to publicize anything they are not prepared to provide. The juvenile firesetter program will lose credibility quickly if the program staff say they can provide prompt assessment and education to firesetters and then place juveniles on waiting lists because they do not have adequate staff.

Strategies

Pamphlets, brochures, and posters. At a minimum, juvenile firesetter programs should develop a simple brochure to describe the program and provide parents and other members of the community with a telephone number to call for additional information. For brochures, pamphlets, and posters, the old adage "less is more" applies. The materials should be simple, with one or two major messages. These materials should briefly highlight the juvenile firesetter program's services and provide a contact for individuals to call. The juvenile firesetter program staff should consider soliciting funds, services (e.g., printing), or in-kind contributions from local businesses to defray the cost of design, production, and mailing. The coordinating council described in the Program Management component may help the juvenile firesetter program with fund-raising for these types of public relations activities.

The brochures can be distributed through the schools, local Parent/Teacher Associations, pre-schools, day-care centers, and pediatricians' offices. Stores may allow the brochure or poster to be displayed in a store window or cashier's desk. Brochures or pamphlets should also be sent to all community organizations, service organizations, hospitals, physicians, and government agencies that work with juveniles.

Newspaper, TV, and radio exposure. The most effective way to publicize a juvenile firesetter program is through local news media exposure. For example, Columbus, Ohio's juvenile firesetter program was suffering because the public was unaware of its existence. With the help of the local television news media and newspapers, the fire department was able inform the public about the problem of juvenile firesetting and the services offered by the program. Program staff gave interviews about the local juvenile firesetter problem and explained how the community could use the Columbus Juvenile Firesetter Program.

The juvenile firesetter program staff cannot wait for the media to come to them. They must go to local newspapers, radio stations, and television stations. If the juvenile firesetter program wants to use the local media, the program coordinator or spokesperson will have to call or meet with news reporters, assignment editors, and local news show producers. It is the spokesperson's job to "sell" the story to local media, explaining the importance of getting information about the juvenile firesetter program to the community. The spokesperson needs to have a clear understanding of the message the program wants to convey to the public and be able to convey that message to the local media.

The spokesperson should have three or four key pieces of information to convey. Examples may include messages such as, 1) the majority of firesetters are curious children who need education, 2) the key to providing services to firesetters, whether curious or troubled, is identification and assessment, 3) children playing with fire is a very real and dangerous problem, 4) parents should not be afraid to seek assistance if they suspect that their children are playing with fire, or 5) the juvenile firesetter program is designed to provide assessment, education and referral for firesetters. Local communities may have messages that apply to their specific jurisdiction. A second key requirement is knowing the target audience. A message targeting parents may be different than a message targeting community agencies. The issue of target audience will be discuss further in the section on Public Service Announcements.

Juvenile firesetter programs should consider writing brief fact sheets and press releases which can be made available to the local media. Fact sheets can be used to give the media background information about the juvenile firesetter program. Fact sheets are usually brief and can be updated as necessary. A press release is a brief (one page) announcement of a newsworthy story or event. The release gives the important information about the event to the media. Every release should have the name, address and telephone number of the juvenile firesetter program.

An excellent resource on how to work with the media was written and published by the National Crime Prevention Council (NCPC). The book, Ink and Airtime, provides ideas and step-by-step guidelines on how to write press releases, fact sheets and articles. The book also tells readers how to get their information on radio and television and how to systematically develop a media campaign.

Juvenile firesetter programs can also benefit from the information in media kits developed by the U.S. Fire Administration (USFA) and the National Fire Protection Association (NFPA). Both media kits contain information on the nature and extent of juvenile firesetting which can be used to educate communities about the problem of juvenile firesetting.

One way to get media attention is when a juvenile firesetting incident has occurred. Parental interest and media awareness are heightened after such an event. A description of the program and its services can be use as a sidebar to the story about the incident. In addition, if the juvenile firesetter program staff have identified themselves to the media, they may be interviewed and asked to give an expert opinion about juvenile firesetting. This is again an excellent opportunity to discuss the juvenile firesetter program. The important thing to remember is that no juvenile firesetting incident should be reported in the media without also mentioning the juvenile firesetter program.

In addition to newspapers, radio, and television, juvenile firesetter programs can also use community newsletters or magazines, newsletters of major corporations, and university and college newspapers to publicize the program. The program staff can write short articles about the juvenile firesetter problem and the steps the program has taken to alleviate the problem. Program staff can then meet with the editors of the newsletters and magazines to discuss the articles. These types of publications are designed to serve the community and highlight community programs and activities and can be an excellent way to educate the community about the juvenile firesetter program.

Public Service Announcements. Public service announcements (PSAs) can also be used to inform the community about the juvenile firesetter program. They have the potential to reach a wide audience. PSAs provide information about a problem or program without trying to sell a product. One of the greatest advantages of PSAs is that the radio and television time are donated by the media. Competition for media time and space, however, is very tough and stations are cutting back on the amount of airtime they are willing to devote to PSAs. PSAs must, therefore, be well thought out and creative.

Several fire departments and the National Fire Protection Agency (NFPA) have developed "generic" or open-format PSAs. These are PSAs that describe the problem of juvenile firesetting in general, but allow the local program to "customize" the PSA by leaving space at the end for information about how to contact the local juvenile firesetter program. The PSA developed by the NFPA entitled, "Got a light, keep it out of sight," can be ordered through local NFPA representatives. The Phoenix Fire Department also has an open-format PSA developed by Fire-Pal. The Phoenix Fire Department has made the PSA available to local juvenile firesetter programs for a reproduction fee.

Juvenile firesetter programs may want to develop their own PSAs. Before developing a PSA, the juvenile firesetter program staff must decide who they want to reach -- their target audience -- and the best way to reach them. Two of the largest, relatively untapped, sources of referrals are parents and school personnel. PSA should be designed to capture the attention and support of these two groups.

The juvenile tiresetter program staff will need to decide on the content of their message. PSAs are usually short, 15-30 second ads that focus on a specific message. The content of the PSA message will vary according to a number of different criteria, including target audience (parents, children, teachers, etc.), nature of the juvenile firesetting problem in the community, and the goal of the PSA (education, referral to the program, etc.). Some PSAs, targeted toward parents, describe misconceptions about juvenile firesetters. One such misconception is that they are "bad" kids or that they have deep-rooted psychological problems. Although some juvenile firesetters are troubled and need counseling, the majority are young children who need fire safety education. Other PSAs are used to inform the public that juvenile firesetting is a real and deadly problem that, in many cases, can be avoided. Still others may address kids and warn them about the dangers of playing with matches and lighters. Regardless of the message, the PSA should give the audience a specific name and telephone number to contact for more information.

Unfortunately, although the media donates PSA time and space, developing a PSA is not a low cost venture. Programs with limited funds will need to look to the community for funds or services. Companies may be able to donate paper, tapes, personnel, video equipment or other valuable materials in lieu of money. Programs unfamiliar with producing audio and video tapes may want to consider using PSAs which have already been developed. The vehicle used to promote the juvenile firesetter program (radio or television) will depend largely on the amount of resources available.

If the resources and expertise are available, the juvenile firesetter program will still have to compete with other agencies for the media time and space. Program personnel should address this problem directly by going to local newspapers and radio and television stations and meeting with the public service staff. The juvenile firesetter program director or another staff member will need to explain the severity of the problem and the importance of eliciting community support for the juvenile firesetter program. Ink and Airtime advises program staff to go to these meetings armed with all of the information available, including local and national statistics, evidence of program success, and endorsements from prominent members of the community.

Speakers bureaus, hot lines, and other services. The juvenile firesetter program or the coordinating council can establish other services to promote the program. For example, juvenile firesetter programs in Columbus, Ohio and San Jose, California have established speakers bureaus. These bureaus are comprised of individuals who have expertise in one or more areas of fire safety and prevention. These individuals volunteer their time to speak to community groups, schools, service organizations, and other interested groups. The speakers can provide valuable information and promote the use of the juvenile firesetter program.

The Juvenile Firesetter Prevention Task Force, Inc. in Columbus, Ohio also maintains a Juvenile Firesetter Care Line where parents can receive information and help for their children. Volunteers from the community can be trained to man the hot-line and assist parents.

Partnerships

The nature and extent of the juvenile firesetter program publicity and outreach campaign will be limited to the resources available to the program. Programs with limited money and manpower have formed partnerships with community organizations and local businesses to acquire the necessary services, materials, and funds. The community can offer an unlimited wealth of resources. Corporations may donate money or sponsor specific promotional activities or products. As noted in the Program Management component, the juvenile firesetter program staff should appeal to a corporation's sense of civic mindedness and self-interest when attempting to solicit donations from corporations. Contributing money to better the community is basically good business. The juvenile firesetter program coordinator should also request assistance from individual community members. Individuals with expertise in writing, advertising, audio and visual communications, design, and other skills can be asked to donate their skills. The problem of juvenile firesetting is a *community* problem that cannot be alleviated without the assistance of the community.

Resource List

Ink and Airtime:
National Crime Prevention Council (NCPC)
1700 K. Street, N.W., Second Floor
Washington, DC 20006

Public Service Announcements:

Fire Pal
c/o Phoenix Fire Department
520 West Van Buren
Phoenix, AZ 85003

National Fire Protection Association
1 Batterymarch Park
Quincy, MA 02169
(6 17) 770-3000

Media Kits

"Curious Kids Set Fires"

U.S. Fire Administration
National Fire Academy
16825 South Seton Avenue
Emmitsburg, MD 21727

"Big Fires Start Small"

National Fire Protection Association
1 Batterymarch Park
Quincy, MA 02169
(6 17) 770-3000

Newsletters

"Hot Issues"
State Fire Marshall
4760 Portland Road, N.E.
Salem, OR 97305

CHAPTER 6: MONITORING SYSTEMS

The purpose, content, format, and use of systems for monitoring juvenile firesetters and firesetting incidents are covered in this component. While many juvenile firesetter programs have developed some internal system to monitor their caseloads, others simply maintain individual case files with no systematic way to track cases, determine final dispositions, report to funding agencies, etc. Very few have systems capable of being used for evaluation purposes. As described in this component, simple monitoring systems are recommended for all juvenile firesetting programs regardless of size. They need not be elaborate, expensive, semi-comprehensible computerized systems; both manual and simplified computer systems can be perfectly adequate for careful monitoring.

Purpose

Monitoring systems serve different purposes, depending on the information they contain and the uses to which they are put. At the most elemental level, a management information system is needed for case tracking, caseload analysis, and reporting of program operations and results. A Management Information System (MIS) should include case characteristics of the firesetter and the firesetting incident, services rendered, dates of key events, and the final disposition of the case. It is used as a management tool to monitor individual cases, determining the status of each case at any given point and ensuring that needed treatment has been completed. An MIS provides the means for summarizing and analyzing the program's caseload (the number of cases handled, case type, firesetter characteristics, number and type of services rendered, etc.), tracking and reporting the number and type of program activities (presentations given, etc.), and providing data for annual reports, evaluations, and funding agencies. Most juvenile firesetter programs maintain some version of an MIS, or at least have the basic ingredients (such as case records) for the making of one.

Extending the MIS to include recidivism and other follow-up data provides the basic building blocks for an evaluation system. An evaluation system would contain all of the information above plus follow-up data on firesetting recidivism and other problems such as delinquency, school or family problems, etc. The evaluation system is an extension of the MIS, rather than a separate system. Much of the data in such an evaluation system may come from the program's routine follow-up contacts with families of firesetters and the referral agencies to which they are referred. It provides the basic data needed for self-evaluation and program monitoring, as well as those needed for an independent evaluation of the program. Some juvenile firesetter programs have routinized systems for tracking recidivism and judging the effectiveness of program efforts, and will be used to illustrate the purposes and use of an evaluation system.

The third type of monitoring system suggested for juvenile firesetter programs is an incidence reporting system. The purpose of an incidence reporting system is to record basic information on all known juvenile firesetting incidents, whether or not the tiresetter is identified and handled by the juvenile firesetter program. This system would provide the basic data needed to monitor jurisdiction-wide rates ofjuvenile arson and firesetting and gauge the effectiveness of education and outreach efforts of the program. The incidence reporting system would be an extension of the routine records kept by fire service officials, and ideally, would include firesetting incidents that have not come to the attention of law enforcement and fire officials.

As noted earlier, we believe that, at a minimum, the juvenile firesetter program should be aware of every juvenile-set fire in its jurisdiction. Sometimes schools and other agencies choose to handle a juvenile firesetter internally without contacting the juvenile firesetter program, particular if the fire did not warrant fire suppression. It is very important for the juvenile firesetter program to have that information, because if a child ultimately does set a fire that warrants the attention of the fire department, it is likely that the fire service will treat that individual case differently if they believe it is a first offense versus if they know that the child in question has been lighting numerous fires in school. In order to provide the child with the services he or she needs, the juvenile firesetter program needs to be aware of the child's firesetting history. Developing strong referral networks should facilitate obtaining the necessary information.

Central Elements of the Monitoring Systems

The case information and other data to be kept in each of the proposed three systems are described in this section. The development, form, and use of these systems -- data collection issues, whether systems should be manual or computerized, which agency should maintain the system, analysis and reporting, etc. -- are described in the following section, "System Development and Use".

Management Information System. The data to be included in the Management Information System are drawn from the individual case files, primarily from intake, screening, and assessment instruments, and from other program records (perhaps newly created for this purpose). There are four categories of data included in an MIS:

I.   Case characteristics

    Source of referral
    b.   Age, sex, race, family status of firesetter
    c.   Details of the firesetting incident--motive, presence of others, location of fire, materials used, damage estimate, injuries, deaths
    d.   Past firesetting incidents
    e.   Initial assessment after screening (e.g., little, definite, or extreme risk)

II. Services rendered

    dates, content, and length of educational sessions; dates, purposes, and agencies of referral(s); number and type of counseling sessions; details of other services (restitution, community service, etc.)

III. Case disposition

    a.   Dates and outcomes of all services rendered, gathered through routine reporting by all cooperating agencies or direct follow-up
    b.   Status of case in criminal justice system

IV.   Program Activities

    a.   Education/prevention activities, school-based or community or other -- type, number, attendance, content
    b.   Training for others in the field -- type, curriculum, number trained
            Resource/materials development
    c.   Other -- media coverage, Task Force participation, etc.

The first three categories, case characteristics, services rendered, and case disposition information, are the most important elements of the Management Information System. The data will be as accurate and complete as the individual case files and other program records. Each case should have a case file which would contain intake forms, screening instruments, and disposition information in each case file.

Evaluation system. Data for an evaluation system requires follow-up activities with police, fire, prosecution, courts, and probation agencies; schools; parents; social service agencies; and private treatment facilities. Data collection procedures are discussed in the following section. The following information on all cases handled is to be included in the evaluation system:

Firesetting recidivism -- information on any further firesetting incidents.

Delinquency -- any and all acts of vandalism, stealing, etc.

School problems -- truancy, chronic tardiness, disciplinary problems, academic and behavioral problems, etc.

Family/home problems -- running away, lack of parental control, etc.

Personal and interpersonal problems -- emotional and behavioral problems, poor peer relationships, etc.

Incidence reporting system. Like the evaluation system, the incidence reporting system requires information from a variety of sources, although the fire department is clearly the primary source. The incidence reporting system should cover the jurisdiction of the juvenile firesetter program -- e.g., a city, county, etc. It will include all known or suspected juvenile firesetting incidents, whether or not they are reported to the authorities (data collection is discussed in the following section). The system would include information on:

Firesetting incidents -- date, location, ignition materials used, items/structures ignited, damage estimate, injuries, death, reported or not, reasons for not reporting if known.

Known or suspected firesetters -- age, sex, motive, presence of others, past incidents.

System Development and Use

Management information and evaluation systems. The development and use of the management information and evaluation systems will be covered here under one heading. The evaluation system should be considered simply an extension of the MIS because the data collection, computerization, and other issues are quite similar. The fire department is best equipped to build and maintain these systems.

Cooperation and coordination from all agencies is needed to build and maintain a comprehensive monitoring and evaluation system. The responsibility of each agency in regard to reporting requirements and providing data should be spelled out in interagency agreements. For a developing program, monitoring issues should be identified from the start and built into discussions among involved agencies from the inception of program planning. These issues include confidentiality, policies or regulations that prohibit sharing information, routinized data collection procedures, and resistance to participating in data collection because of time constraints or lack of resources.

Each agency may have confidentiality concerns, at minimum, or statutory regulations that limit the individual information they may share with others. Schools, in particular, will be very concerned about releasing any information on students. Law enforcement agencies, particularly probation and the courts, are severely limited in the extent to which any information on juveniles can be provided to outsiders, and in some jurisdictions, the fire and police departments are considered outsiders. While these concerns may be mitigated if the monitoring system is maintained by fire officials within the juvenile firesetter program, confidentiality protections should be reviewed and safeguarded. Access to records should be controlled whether the system is manual or computerized, and only grouped results should be reported outside of the program.

Each agency involved in the juvenile firesetter program must make a commitment to inform the program about particular events concerning youth in the program. The data collection effort should not be burdensome; simple reporting forms can be developed to facilitate case tracking and disposition. In one program, for example, counseling agencies submit monthly reports on clients referred to them by the juvenile firesetter program and Child Protective Services reports quarterly on the number of family sessions held and progress made. When a family terminates counseling, the entire case file kept by the counselor is returned to the juvenile firesetter program. In another program, the program follows up with both the parents and the referral agencies within a month or two after referral to confirm that the recommended contact has occurred.

The management information and evaluation system may be kept manually, but since personal computers have become increasingly prevalent in the workplace, computerization is advised. A manual system may suit a small program perfectly, if its caseload is not large and its reporting requirements are small. Simple logs, carefully organized and kept up to date, will provide a small program with basic information very quickly. Computerization is needed when either the caseload is too large to handle summary computations easily and accurately or when reporting requirements are frequent and/or detailed, making interim computations and status reports cumbersome to produce. When a program reaches somewhere between 75 and 100 cases per year, computerization is probably warranted.

If operating on a manual system, key information from case files should be placed on monthly activity logs that enable summaries to be easily calculated. For example, the sample log on the following page will tell you at a glance the number of cases handled in July, their referral sources, and initial intervention steps. With minor calculations, the average age and other information about the firesetter and firesetting incident can be summarized. A program developing a manual MIS should decide what information is to be kept on logs, after reviewing their management and reporting needs. Perhaps two logs will be needed, one to describe basic case information (dates, referral sources, individual and incident characteristics) and one to record referral, intervention, and disposition data. A log should be used as a "tickler system", enabling program staff to quickly view the status of a case and monitor it for the delivery of intervention services.

In a computerized system, information from case records would be entered directly into a computer using a database management program (e.g., Dbase). Simple queries on a case by case basis can be made through the database program, such as the date of referral to the program, and many database programs enable more complex queries to be made easily, such as the number of cases referred to the Community Mental Health Center. Tables, summary statistics, and routine reports can be produced by programming through the database program. Statistical packages such as SPSS are probably not needed for monitoring purposes, although their statistical capabilities may be helpful in producing specific information needed by a juvenile firesetter program.

One advantage of a computerized system is that it provides a basic database from which information can be drawn, sliced anyway the program desires. For example, manually kept logs can provide a program with a running total of the years' caseload. But if new questions or needs arise -- to look at referral sources during a given quarter or investigate

July 1993

Name	Age	Sex	Fire Location	Property damage amount	Injury Y/N	Screened Y/N	Intervention	Referred from	Referred to

whether kids 13 or over have caused more serious fires than those under 13, for example -- the hand tallies can become burdensome and inaccurate. Such information would be at your fingertips in a computerized system. To maintain an MIS capable of providing a full picture ofjuvenile firesetting in a given jurisdiction, computerization is needed. A computerized database can contain much more information than a manual system (the results of screening tests and details of the firesetting incident, for example) and therefore answer much more complex questions (such as do second-time firesetters improve more with inpatient or outpatient treatment). Without a computerized system, these questions require the hand-culling of individual case files. On the other hand, computerized systems require more care. Data must be entered on a timely basis and one must know how to get information out of it without expending substantial time in training or programming.

A generic computer system could be developed by a central agency (such as the U.S. Fire Administration/FEMA) at relatively low cost and tailored to each program as needed. Alternatively, a juvenile firesetter program might hire local experts (or find volunteers among local computerniks) to develop a computer program. We feel most juvenile firesetter programs will have the in-house expertise needed to enter data and operate the system on an ongoing basis. It should not be a time-consuming job; rather only an hour or two a week should keep the system current.

The analysis produced via the computerized system or summarized manually in regard to the management information system should enable a program to answer questions such as the following:

1. How many cases have been handled this year relative to last year?
2. What are the individual and family characteristics of the juveniles handled?
3. What are the characteristics of the fires set by the juveniles handled by the program?
4. Which referral agencies are used the most?
5. How long, on the average, are juveniles and families in treatment?

To extend the MIS to become an evaluation system, follow-up activities must take place with a number of key agencies to determine the long-term effectiveness of the intervention strategies in terms of recidivism. For evaluation purposes, a program needs to know, minimally, of any recurrence of firesetting behavior, and should want to know about juvenile delinquency, continued problems at school or home, etc. Quarterly contacts should be made with the family and key agencies for a year or two after the precipitating incident to inquire about recidivism and related problems. In the Houston program, cross-reference checks are made among participating agencies to look for recidivists and the program director makes monthly phone calls to the family for a year to check on the juvenile's progress. Other programs have formally conducted surveys of families to explore recidivism issues and what the family felt about the juvenile firesetter program and the referral services that may have been offered. Examples of these surveys can be found in the *Guidelines for Implementation*

The key agencies include the police and fire departments, courts and probation, schools, parents, social service agencies, and public treatment facilities. The follow-up may consist of routine reporting or periodic telephone calls to determine if the agency has had any further contact with the juvenile and, if so, for what reasons. Parents are probably the best single source of follow-up information, if sufficient rapport has been built to enable the parents to report any additional delinquent behaviors or other problems. Telephone contact should be made with the parents rather than sending an impersonal form.

These recidivism data should be added to the computerized database or manual logs as they are gathered. Together with the MIS data, this information forms the basis for a comprehensive evaluation. The information is obviously valuable to the program, to assess its own effectiveness and effectiveness of participating agencies. An independent evaluator will want to verify the information and collect more detailed information on treatments and outcomes, but the MIS will provide the basic building blocks for an outside evaluation. Finally, the MIS data are easily available when preparing annual reports, proposals, news releases, etc.

Incidence reporting system. Incidence reporting systems, as discussed previously, are valuable for analyzing the full problem ofjuvenile firesetting and determining where services are needed and where services (education, particularly) have been effective. Since fire departments will have the basic systems in place needed to maintain an incidence reporting system, the real challenge is in data collection. In too many jurisdictions, there is no means to identify fires set by juveniles versus those set by others.

A juvenile firesetting incidence reporting system should contain fire and individual information as previously presented. The data should be gathered via existing records or new forms developed for this purpose, from all fire departments covering jurisdictions of interest depending on the areas served by the juvenile firesetter program. The Portland, Oregon program is building a statewide database on juvenile firesetters. Portland has also conducted a risk analysis of the city to identify high-risk areas for juvenile firesetters and implement education/intervention strategies as appropriate. The state of New York, in conjunction with the Rochester program, is also developing a statewide computer system.

In addition to gathering and analyzing reported juvenile firesetting incidents, methods to assess the incidence of unreported fires are needed. Several options are available. One way is to identify and survey organizational entities (primarily schools and parents organizations) that record firesetting incidents that are small and not reported to the fire department.

Another, more basic assessment of the juvenile firesetting problem is to survey youth directly to gather information on their firesetting behavior. Juvenile firesetting is substantially underreported, and many youth set fires that never come to the attention of parents or authorities. Anonymous surveys of students in the schools (as conducted by the Rochester program) are probably the best single source of information on juvenile firesetting incidents as well as fireplay activities that do not result in actual fires. Strict anonymity must be upheld for truthful self-reports to result. This type of survey will provide information on the full extent of the juvenile firesetting problem in a jurisdiction and is as valuable as reported fire statistics.

Because of the volume of data and need for summary statistics, the incidence reporting system should be computerized. In many departments, the creation of this system will be relatively easy. Fire incidence reports that are routinely computerized may be sorted to reflect just the juvenile problem.

CHAPTER 7: DEVELOPING RELATIONSHIPS WITH THE JUSTICE SYSTEM

The juvenile firesetter program needs to develop relationships with all of the key agencies that work with juvenile firesetters. As noted in the Referral Mechanisms component, these agencies include police, social services, schools, mental health, and the juvenile justice system. It is the relationship with the juvenile justice system which will be the focus of this component. Too often juvenile firesetters are referred to the juvenile justice system and never come to the attention of the juvenile firesetter program. Juvenile firesetter programs need to be aware of all incidents of firesetting so that no juvenile firesetter falls through the cracks. The juvenile firesetter program is in a unique position of being able to assess all firesetters and track them through referral agencies. In addition the juvenile firesetter program can be a resource for probation, juvenile court, and correctional facilities. The juvenile firesetter program's potential as a resource to the juvenile justice system, however, is based on strength of relationship between the program and the justice agencies.

Purpose

The purpose of this component is to provide information to assist programs in developing effective relationships between the juvenile firesetter program and the justice system. Specific objectives of this component are the following:

. to help the juvenile firesetter program identify deliquent firesetter;

. to develop working relationships between the juvenile firesetter program and specific agencies of the criminal justice system including probation, family court, prosecutor's office, and juvenile court; and

. to help the juvenile firesetter program better assist delinquent firesetters in residential correctional facilities.

Model approaches that have been used in three cities - Rochester, NY, Charlotte, NC, and Portland, OR - are summarized below. Recognizing that each program will differ in such areas as state law, program structure and manpower, programs should review these models and consider the procedures most appropriate to their particular program and jurisdiction.

o   **Rochester, New York**

All fires set by juveniles are investigated by fire investigators assigned to the Fire Related Youth (FRY) Program. Investigators approach every case of juvenile firesetting as a criminal investigation. After collecting information about the case, the investigators meet with the youth and provide fire safety education. In Rochester, all juvenile firesetters come to the attention of the FRY program prior to being referred to other agencies, including juvenile justice. Cases may be referred to prosecution as a last resort to get services to children in need or when the child has engaged in numerous other delinquent activities. The FRY investigators work very closely with the Probation Office and the Presentment Agency. The Probation Intake Unit will handle the case first. Intake Unit staff will decide whether the case can be "adjusted" without going to court or whether the case will be referred for prosecution. If adjustment is being considered, the probation department will consider the youth's risks and strength. Probation staff may refer the juvenile to a mental health or social service facility. Cases referred to prosecution are petitioned through the Presentment Agency to Family Court. The presentment attorneys work very closely with the FRY investigators and rarely lose a case sent to prosecution. Before a case gets to court, the judge assumes that every effort has been made to keep the youth out of court. If the judge finds that there is enough evidence to justify the charge, s/he will ask Probation to conduct a family evaluation and make a recommendation to the court. Often the judge will also consider the FRY investigator's recommendation.

o   **Charlotte, North Carolina**

All arson or suspicious fires are investigated by the Arson Task Force. If a juvenile is suspected, the Task Force members often try to persuade the youth to confess to setting the fires. How the case proceeds often depends on whether the juvenile confesses. If the juvenile does not confess, the case is referred to the District Attorney's Office for prosecution. If the juvenile does confess, the Task Force then decides whether to proceed with prosecution or refer the juvenile directly to the juvenile firesetter program which is housed in another division of the fire service. The decision is based on a number of considerations including whether the incident is a first offense and whether the youth destroyed another's property. If the youth is referred to court, the case is referred to a Court Intake Counselor. The intake officer will meet with the parents and the youth and decide whether the petition for prosecution is warranted. The intake officer can recommend a deferred sentence under the condition that the youth participate in the juvenile firesetter program. If the case goes to court, the youth is interviewed by a court counselor who makes a recommendation to the court. The court counselor can also recommend that the youth participate in the juvenile firesetter program. If such a recommendation is accepted by the court, it then becomes a court order. The judge may also court order the youth to other agencies or facilities.

o   **Portland, Oregon**

In Portland, Oregon, all juvenile firesetters are reported to and, investigated by, the Portland Fire Bureau. In 1986, the Fire Bureau developed a program to reduce the incidence of juvenile firesetting. Juveniles apprehended for fire related offenses may be referred *directly* to the juvenile firesetter program or they may be referred to the program via the juvenile justice system. Those referred to the justice system are more likely to be the older juvenile who has been 1) involved in a more serious incident, 2) identified as a "troubled firesetter," or 3) identified as a repeat offender. If a youth is referred to juvenile court, the case is assigned to an intake probation officer. The intake officer will review all of the records and make a decision to close the case; divert the case to the juvenile firesetter program, social service agency, mental health professional, or another diversion program; or refer the case to the District Attorney's Office. If the district attorney chooses to prosecute the case, a petition of charges is filed with the juvenile court. At this point the case is assigned to an adjudication officer, who prepares the case summary and recommendations. If the judge finds that a crime was committed, s/he must decide whether to sentence the juvenile to a correctional facility, mental health facility, or place the juvenile on probation. As a condition of probation, the juvenile may be court ordered to attend fire safety education through the juvenile firesetter program, or participate in mental health counseling.

Relationships with the Probation Department

Within the justice system, a representative of the probation department (intake unit) is usually the first person to encounter the juvenile firesetter. Therefore, the juvenile firesetter program must inform and educate the probation department, especially those assigned to the intake unit, about the program. For example, a representative from the juvenile firesetter program should make an in-service education presentation to the staff of the probation department.

The staff of the probation department should receive periodic updates, fact sheets, newsletters or yearly updates as to the status of the juvenile firesetter program. Prepared by the juvenile firesetter program staff, these communications can contain statistics, case studies, intervention techniques, list of placement facilities, referral methods, etc. The updates are designed to keep the probation department abreast of what the juvenile firesetter program is doing.

The juvenile firesetter program should plan and coordinate a procedure by which the probation department refers all juvenile firesetters to the program for an evaluation if such an evaluation is warranted. This process will ensure that all juveniles are identified and evaluated and offered educational intervention, if appropriate. One way to plan and coordinate such a procedure which has been used is some jurisdictions would be for the probation department to assign a particular probation officer (most likely in the intake unit) to handle all cases involving juvenile firesetters. That intake officer would be able to work closely with the juvenile firesetter program staff.

In addition, a representative from the juvenile firesetter program routinely should be present at all conferences concerning the treatment and/or placement of a juvenile firesetter. Input from the juvenile firesetter program will be invaluable in discussions with child protection agencies, mental health agencies, correctional facilities, and representatives from community placements.

Relationship with the Law Enforcement, Legal and Judicial Community

The members of the law enforcement, legal (prosecutive and defense), and the judicial community must be aware of, and educated about, the juvenile firesetter program. Certainly, the juvenile firesetter program can be an invaluable referral source for the district attorney's office, trial lawyers, and juvenile judges.

Effective methods of informing and educating the members of these professional communities include supplying them with brochures explaining the program, conducting in-service education seminars, and sending fact sheers, periodic newsletters, and annual reports about the activities of the program. These methods will not only inform and educate, but will also continue to enhance the professional image of the juvenile firesetter program. Such an image is imperative if the professional community is to utilize the services of the juvenile firesetter program.

Relationships with the Juvenile Correctional Institutions

Some juvenile tiresetters will be placed in juvenile correctional institutions for rehabilitation. The juvenile firesetter program can also educate the various correctional institutions about the existence and the contents of the program. Similar relations should be fostered with the correctional institutions as with the probation department.

For example, the juvenile tiresetter program should be aware that a juvenile firesetter is being held at a particular correctional institution. Also, the juvenile firesetter program, once it is aware that a juvenile firesetter is to be admitted to a correctional institution, should inform the institution that the program has evaluated and/or educated the juvenile. A dual waiver, which is signed by the juvenile firesetter's parent or guardian, would allow the juvenile firesetter program to share information they may have about the juvenile with the correctional facility and allow the facility to share information with the program.

The juvenile firesetter program should provide periodic in-service education programs to appropriate staff of the correctional facilities, many of whom are likely to hold inaccurate perceptions of the juvenile tiresetter. For example, the overwhelming majority of correctional facilities, as well as community placements such as halfway houses, believe that the juvenile firesetter is a highly dangerous individual. They perceive the juvenile firesetter as one who is always on the verge of acting out and starting a fire. In actuality, the juvenile firesetter is less likely to act out by starting a fire once he/she is placed in a structured environment and away from the psychological and sociological factors that helped produce the original firesetting behaviors.

The juvenile firesetter program should maintain an open line of communication with the correctional facilities. Correctional facilities rarely maintain specific treatment programs for juvenile firesetters. One such program, is operated by the Upper Arlington, Ohio Juvenile Firesetter Program. The Upper Arlington program offers a 12 week educational program to juveniles incarcerated for arson. Juvenile firesetter programs should encourage and participate in the development of similar programs for juveniles. If a structured program is not possible within the correctional facility, then juvenile firesetter programs should make its staff readily available to the staff of the correctional facility to establish individual treatment plans for specific cases.

www.ingramcontent.com/pod-product-compliance
Lightning Source LLC
Chambersburg PA
CBHW081620170526
45166CB00009B/3052